Footbridge Vibration Design

Footbridge Vibration Design

Editors

Elsa Caetano & Álvaro Cunha
University of Porto, Porto, Portugal

Joel Raoul
Setra, Bagneux Cedex, France

Wasoodev Hoorpah
M.I.O., Paris, France

CRC Press
Taylor & Francis Group
Boca Raton London New York

CRC Press is an imprint of the
Taylor & Francis Group, an **informa** business

Cover photo: Maquette of design proposal for a footbridge over the river Douro
in Porto, Portugal, by A. Adão da Fonseca

CRC Press
Taylor & Francis Group
6000 Broken Sound Parkway NW, Suite 300
Boca Raton, FL 33487-2742

First issued in paperback 2017

CRC Press/Balkema is an imprint of the Taylor & Francis Group, an informa business

© 2009 by Taylor & Francis Group, LLC

Typeset by Vikatan Publishing Solutions (P) Ltd., Chennai, India

Published by: CRC Press/Balkema
                P.O. Box 447, 2300 AK Leiden, The Netherlands
                e-mail: Pub.NL@taylorandfrancis.com
                www.crcpress.com – www.taylorandfrancis.co.uk – www.balkema.nl

Library of Congress Cataloging-in-Publication Data

Footbridge vibration design / editors, Elsa Caetano ... [et al.].
      p. cm.
  Papers from a workshop at the 3rd International Footbridge Conference, July 2008,
in Porto, Portugal.
  Includes bibliographical references and index.
  ISBN 978-0-415-49866-1 (hardcover : alk. paper) – ISBN 978-0-203-87466-0
(e-book) 1. Footbridges–Design and construction–Congresses. 2. Bridges–
Vibration–Congresses. I. Caetano, Elsa de Sá, 1965- II. International Footbridge
Conference (3rd : 2008 : Porto, Portugal) III. Title.

TG428.F66 2009
624.2'5–dc22

2009006421

ISBN 13: 978-0-415-49866-1 (hbk)
ISBN 13: 978-1-138-11328-2 (pbk)

Visit the Taylor & Francis Web site at
http://www.taylorandfrancis.com

and the CRC Press Web site at
http://www.crcpress.com

Contents

List of authors

James Brownjohn
University of Sheffield, Sheffield, UK

Elsa Caetano
University of Porto, FEUP, Porto, Portugal

Xavier Cespedes
SETEC-TPI, Paris, France

Pascal Charles
SETRA, Bagneux, France

Álvaro Cunha
University of Porto, FEUP, Porto, Portugal

Philippe Duflot
Taylor Devices Europe, Brussels, Belgium

Markus Feldmann
RWTH Aachen University, Aachen, Germany

Olivier Flamand
CSTB, Nantes, France

Christoph Heinemeyer
RWTH Aachen University, Aachen, Germany

Wasoodev Hoorpah
MIO, Paris, France

Angus Low
Arup, London, UK

Filipe Magalhães
University of Porto, FEUP, Porto, Portugal

Christian Meinhardt
Gerb Vibration Control Systems, Essen, Germany

Carlos Moutinho
University of Porto, FEUP, Porto, Portugal

Aleksandar Pavic
University of Sheffield, Sheffield, UK

Doug Taylor
Taylor Devices Inc, North Tonawanda, USA

Stana Zivanovic
University of Sheffield, Sheffield, UK

Theodor Zoli
HNTB, New York, USA

Krzysztof Zoltowski
Gdansk University of Technology, Gdansk, Poland

Preface

The current trend for construction of footbridges marked by increased lightness and slenderness has brought new requirements into design. In effect, despite the small loads associated with pedestrian action, two episodes of large vibrations in two landmark footbridges, the Millennium Bridge in London, and the Solférino Bridge in Paris, have shown the importance of dynamic effects in these long bridges and the need to develop methodologies for their assessment and control.

In the last decade an intensive research has been done on the topic of human-induced vibrations in footbridges. This research is now getting more accredited by the establishment of design guidelines and recommendations.

This collection of 10 invited contributions by eminent experts results from the workshop "Footbridge vibration design: worldwide experience", held during the 3rd International Footbridge Conference, July 2008, in Porto, Portugal, and organised by the Faculty of Engineering of the University of Porto in conjunction with the Technical Advisory Bureau for Steel Users ConstruirAcier.

This volume aims to transfer current knowledge of footbridge dynamic design in an applied, practical way. It presents new design approaches and the most recent numerical, experimental tools for assessing and controlling pedestrian effects, thoroughly supported by practical cases. The book's focus is on:

1 Recommendations, guidelines, codes, design tools
2 Practical experience – case studies

In the first part, guidelines and codes, as well as experimental and numerical tools to assist design, are presented. In the second part, practical experience is unveiled with particular focus on the design and control of footbridge vibrations.

The editors convey their sincere thanks to all the authors for their contributions, expecting that the book will be a valuable tool for Civil and Mechanical engineers involved with the design and research of footbridges. The Portuguese Science Foundation (FCT) and ConstruirAcier/ Made of Steel are acknowledged for their support as sponsors of the Footbridge 2008 Workshop which originated the contents of this book.

Elsa Caetano
Porto, January 2009

Part I

Recommendations, guidelines, codes and design tools

Research performed at European level in the topic of footbridge dynamics led to the recent publication of Recommendations and Guidelines. These documents introduce similar design concepts in which the designer and client agree on the specification of a set of relevant design situations, taking into account different traffic situations and individual demands on vibration comfort.

The paper developed by Christoph Heinemeyer and Markus Feldmann presents the design guide methodology developed within the European Project SYNPEX (Advanced Load Models for Synchronous Pedestrian Excitation and Optimised Design Guidelines for Steel Footbridges, RFS–CR 03019, 2004–2007.

Pascal Charles provides an example of the application of the so-called French guidelines (Guide méthodologique passerelles piétonnes/Technical guide Footbridges: Assessment of vibrational behaviour of footbridges under pedestrian loading, Setra, 2006) to an existing footbridge, discussing different possibilities to improve the dynamic behaviour.

With a design perspective Angus Low discusses how dynamic requirements – both wind and human effects – govern the design of long-span footbridges. In a number of recent case histories he follows the story of lateral pedestrian excitation, and how it affects design.

Referring to numerical available tools for dynamic design, the contribution of Krzysztof Zoltwoski discusses modelling aspects, including structural models and dynamic actions associated with human loads.

Given the importance of the accurate dynamic characterisation of lively footbridges, particularly when control measures are expected, Álvaro Cunha, Elsa Caetano, Carlos Moutinho and Filipe Magalhães discuss the usefulness of several types of dynamic testing tools in the support of design and maintenance of footbridges.

Chapter 1

European design guide for footbridge vibration

Christoph Heinemeyer & Markus Feldmann
RWTH Aachen University, Aachen, Germany

SUMMARY

Increasing vibration problems encountered in the last few years show that footbridges should no longer be designed for static loads only but also for the dynamic actions and vibration behaviour of the footbridge due to pedestrian loading.

For this reason European research has been performed to come up with a design concept for footbridges that takes into account different traffic situations and individual demands on vibration comfort. So the elaborated guideline considers different types of pedestrian traffic and the traffic density which can greatly influence comfort requirements of the bridge and the dynamic behaviour. It is important to predict the effect of pedestrian traffic on footbridges at the design stage and in the later verification of serviceability in order to guarantee a comfort level for the user. This guide gives recommendations for designer and client to find relevant design situations as well as methods how to prove if vibration requirements are fulfilled.

This paper presents the design guide methodology and gives some background information.

Keywords: Footbridge; dynamics; structural concepts; planning; vibration; design guide; damping; comfort.

1.1 INTRODUCTION

Vibrations of footbridges are an issue of increasing importance in current design practice. More sophisticated bridge types like cable supported footbridges or stress ribbon bridges, increasing spans and more effective construction materials result in lightweight structures, which have a high ratio of live load to dead load. As a result of this trend, many footbridges have become more susceptible to vibrations when subjected to dynamic loads. Besides wind loading the most common dynamic loads on footbridges are the pedestrian induced footfall forces due to walking or jogging.

As a consequence of these lightweight structures, the decrease in stiffness leads to lower natural frequencies with a greater risk of resonance, while the decrease in mass reduces the mass inertia. Resonance occurs if the frequency of the bridge coincides with the frequency of the excitation, e.g. the step frequency of pedestrians. Due to the small mass of slender lightweight structures the mass inertia is much lower and hence the dynamic forces can cause larger amplitudes of response. The more slender constructions become, the more attention must be paid to vibration phenomena.

Pedestrian induced excitation is an important source of dynamic excitation on footbridges. The caused vibrations may occur in vertical and horizontal direction, even torsion of the deck is possible.

Vibrations of footbridges may lead to serviceability problems, as effects on the comfort and emotional reactions of pedestrians might occur, although collapse or even damage due to human induced dynamic forces has occurred very rarely.

In recent years some footbridges were excited laterally by dense pedestrian streams in which pedestrians interacted with the bridge vibration. A self-excited large response was produced and caused discomfort. Footbridges should be designed in such a way that this pedestrian-bridge-interaction phenomenon, also called 'lock-in', does not arise.

In Building codes ([2, 3, 4, 5]) this dynamic problem is considered by giving limits for the natural frequency. This rough assumption restricts pedestrian bridge design. E.g. Slender, lightweight bridges, such as stress ribbon bridges and suspension bridges may not satisfy these requirements.

The presented design according to the guideline elaborated within the SYNPEX project [1] takes into account that pedestrian bridges have different traffic situations that may be more or less relevant for design. One exceptional situation is e.g. the inauguration of a bridge with a very dense traffic that occurs often once in the life of a bridge only.

Thus this guideline gives help to find together with the client relevant traffic situations and to define a related comfort that should be fulfilled under that traffic situation. Traffic class and related comfort criteria are the goal of the bridge design.

The guideline also gives methods how to determine the relevant dynamic bridge characteristics and the bridge acceleration under pedestrian traffic.

This paper concentrates on the design procedure. Further information may be found in the SYNPEX final report [1].

1.2   DESIGN PROCEDURE

The design method aims on the proof of comfort for vertical and horizontal vibration. It does not aim in design for structural integrity or fatigue.

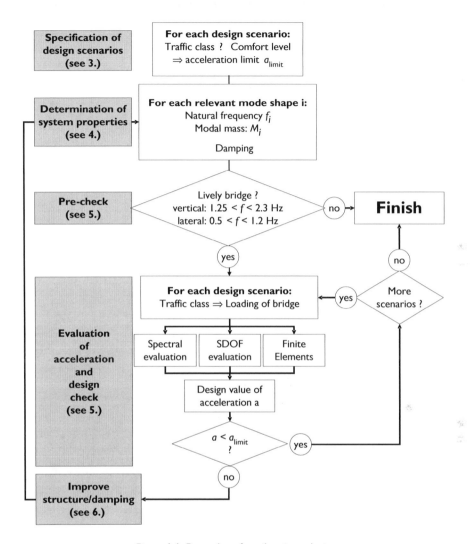

Figure 1.1 Procedure for vibrations design.

By tests and surveys of pedestrians who have just passed a bridge – performed within the SYNPEX project [1] – it has been found that a general definition of comfort is not reasonable but individual definition of comfort criteria should be apllied. Thus the specification of design scenarios is the first design step in the flow chart illustrated in Fig. 1.1. The flow chart also contains the links to the relevant chapters of this paper which include further descriptions.

Another key point in the guideline is the evaluation of expected acceleration of the bridge. Three alternative methods – spectral method, Finite Element Analysis and SDOF method (single degree of freedom) – are presented in chapter 5 and some general information about the determination of system properties are give in chapter 3.

1.3   SPECIFICATION OF DESIGN SCENARIOS

1.3.1   Traffic classes

The expected pedestrian traffic on footbridges depends on various boundary conditions as e.g. location of the bridge:

- in parks or on the country site one expects few people to promenade,
- in a city centre one expects a more persistent weak or dense stream of people, and

Table 1.1   Traffic classes.

Traffic class	Density d (P = Person)	Description	Characteristics
TC 1	group of 15 P; d = 15 P/bl	Very weak traffic	15 single persons (b = width of deck; l = length of deck)
TC 2	d = 0.2 P/m²	Weak traffic:	Comfortable and free walking, Overtaking is possible, Single pedestrians can freely choose pace.
TC 3	d = 0.5 P/m²	Dense traffic:	Significantly dense traffic, Unrestricted walking, Overtaking can intermittently inhibit.
TC 4	d = 1.0 P/m²	Very dense traffic:	Freedom of movement is restricted. Uncomfortable situation, obstructed walking, Overtaking is no longer possible.
TC 5	d = 1.5 P/m²	Exceptional dense traffic	Very dense traffic and unpleasant walking. Crowding begins, one can no longer freely choose pace.

- close to a exhibition hall or a stadium people may cross in intervals and dense or very dense streams.

Table 1.1 gives traffic classes (TC) with appropriate densities of persons, illustrations and descriptions.

1.3.2   Comfort levels

The assessment of the horizontal and vertical footbridge vibration includes many 'soft' aspects such as:

- Number of people walking on the bridge,
- Frequency of use,
- Height above ground,
- Position of human body (Sitting, standing, walking),
- Harmonic or transient excitation characteristics (vibration frequency),
- Exposure time,
- Transparency of the deck pavement and the railing, and
- Expectancy of vibration due to bridge appearance.

The comfort levels for different acceleration ranges of the bridge recommended by the guideline are presented in Table 1.2. In general there are four comfort levels: maximum comfort, medium comfort, minimum comfort and unacceptable discomfort.

It should be clear that passing the bridge is possible for the three acceptable comfort levels C1 to C3.

1.3.3   Specification matrix

As stated above comfort is a soft aspect. For assigning comfort levels to traffic classes the following should be considered:

- Very slender bridges may not be feasible when specifications are too severe – Maximum comfort may not be reached, but depending on further boundary conditions as type of users and location a less restrictive traffic class may be acceptable in design.
- The occurrence of traffic: For unusual traffic situations a minimum comfort may be sufficient; For frequent traffic situations a higher comfort class may be adequate.
- The location of the bridge: Close to hospitals and nursing homes where people pass who are made feeling insecure by vibration a high comfort level might be

Table 1.2 Defined comfort classes with limit acceleration ranges.

Comfort level	Degree of comfort	Acceleration level vertical	Acceleration level horizontal a_{limit}
CL 1	Maximum	<0.50 m/s²	<0.10 m/s²
CL 2	Medium	0.50–1.00 m/s²	0.10–0.30 m/s²
CL 3	Minimum	1.00–2.50 m/s²	0.30–0.80 m/s²
CL 4	Unacceptable discomfort	>2.50 m/s²	>0.80 m/s²

applicable; close to a stadium or exhibition hall medium comfort might may be demanded; in a forest where only few hiker pass the bridge a minimum comfort should be enough.

The assignment of desired comfort levels to traffic classes should be performed together with the client. It is proposed to use a specification matrix as shown in Table 1.3. Table 1.3 is exemplarily filled in for the most frequent situation of footbridges where exceptional dense traffic is not expected and critical persons are exceptional users of the bridges.

1.3.4   Lateral lock-in

Unlike vertical vibrations which are absorbed by legs and joints so that pedestrian streams synchronising with vertical vibrations have not been observed on footbridges, people are much more sensible to lateral vibrations. As for walking the centre of gravity is not only varied vertically but also laterally from one foot to the other, the frequency of the movement of the human centre of gravity is half the walking frequency. If a person walks on a laterally vibrating bridge, he tries to compensate the additional movement of his centre of gravity by swaying with the bridge displacement for lateral stability. This behaviour is intuitive and even small and not perceptible vibrations are assumed to cause an adjustment of the movement of the centre of gravity. This change of the movement of the centre of gravity is accompanied by an adaptation of the walking frequency and a widening of the gait. The person tends to walk with twice the vibration frequency to move his centre of gravity in time with the vibration [9].

The swaying of the body in time with the lateral vibration causes that the lateral ground reaction forces are applied in resonance and the widening of the gait causes an increase in the lateral ground reaction forces. The forces are applied in such a way that they introduce positive energy into the structural system of the bridge (Fig. 1.2). Hence, if a footbridge vibrates slightly in lateral direction and it happens that the pedestrian adjust their walking pattern, then due to this synchronisation effect a low-damped bridge can be excited to large vibrations.

Tests in France [10] on a test rig and on the Passerelle Solferino indicate that a trigger amplitude of 0,1 to 0,15 m/s² exist when the lock-in phenomenon begins.

Table 1.3 Specification matrix.

Comfort level	Traffic class				
	TC 1 Very weak	TC 2 Weak	TC 3 Dense	TC 4 Very dense	TC 5 Exceptional dense
CL 1: Maximum	0	0	0	0	–
CL 2: Medium	X	X	X	0	–
CL 3: Minimum				X	–

Legend: – : Not expected 0; Not demanded X; Demand.

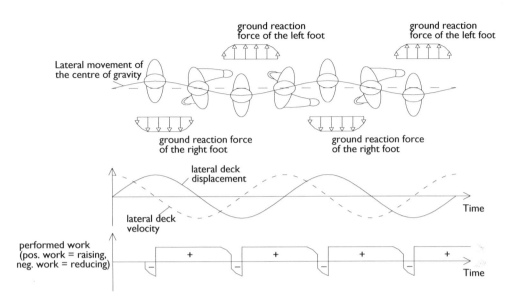

Figure 1.2 Schematic description of synchronous walking.

$$a_{\text{lock-in}} = 0.1 \text{ to } 0.15 \text{ m/s}^2 \tag{1}$$

Experiments within the project SYNPEX [1] on a test rig indicate that single persons walking with a step frequency $f_i \pm 0,2$ Hz tend to synchronise with deck vibration. Fast walking persons are nearly not affected by the vibration of the platform, as the contact time of the feet is short and the walking speed very high. They seem to be less instable than when walking with slow and normal speed.

The lock-in trigger amplitude is expressed in terms of acceleration. Further frequency dependence could exist but has not been detected in measurements.

The definition of comfort classes for horizontal vibration and their determination consider the effects described above.

1.4   DETERMINATION OF SYSTEM PROPERTIES

The determination of the properties of a footbridge depends on the design stage and on the type of structural system.

In early design stages (e.g. pre-design) it may be adequate to apply hand formulas – as given in Table 1.9 for a simply supported beam – if the structure is not too complex.

In later design stages and for complex structures Finite Element Analysis (FEA) is usually applied. The FEA can be performed using beam and/or shell elements. Applying FEA it should be considered if small or large deformations are expected as this has an influence on the structural model. If small deformations are expected hinged connections (as assumed in Ultimate Limit State ULS) may act more like rigid connections. Thus the structural system for dynamic investigations may differ from that for the ULS design.

Table 1.4 Damping ratios ξ according to construction material [11, 12].

Construction type	Minimum ξ	Average ξ
Reinforced concrete	0.80%	1.3%
Prestressed concrete	0.5%	1.0%
Composite steel-concrete	0.30%	0.60%
Steel	0.20%	0.40%
Timber	1.50%	3.0%

Independently from the calculation method damping properties of the structure need to be defined. For the design of footbridges for comfort level Table 1.4 recommends minimum and average damping ratios. Comparable values are proposed by the SETRA/AFGC guideline [11] and by Bachmann and Amman [12].

1.5   EVALUATION OF ACCELERATION

The guideline gives three alternative methods for the evaluation of acceleration:

- Spectral approach: This method has been developed for the analysis of beam bridges. It gives results with minimum calculation effort and is thus very suitable for preliminary design studies.
- SDOF-Method: The SDOF-Method reduces the structural system in regarded mode shape to a *single degree of freedom* system that can be examined easily.
- FEA-Method: With finite elements the most detailed investigation is possible. If the evaluation method presented here can not be applied with FE-program at hand one of the two other methods should be used.

Before starting the evaluation of acceleration it should be checked if it is likely that the footbridge vibrates due to pedestrian traffic. The bridge is lively and acceleration check should be performed if the natural frequency f_i of the regarded mode shape is in the following range:

- For vertical vibration: $1.25 \leq f_i \leq 2.3$ Hz
- For horizontal vibration: $0.5 \leq f_i \leq 1.2$ Hz

The acceleration needs to be determined and checked against the acceleration limit of the appropriate comfort level (Table 1.2) for each design scenario.

1.5.1   Load models

1.5.1.1   General

For the spectral evaluation of acceleration the loading of the bridge is sufficiently defined by the crowd density in P/m² (Persons per square metre). The effects mentioned below have been considered in development of the spectral method.

In the two other evaluation methods the pedestrian induced action is represented as oscillating distributed load $p(t)$. This load is applied in accordance with the mode shape as shown in Fig. 1.3.

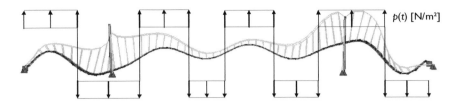

Figure 1.3 Application of harmonic load according to a mode shape configuration.

The harmonic oscillating load is defined by:

$$p(t) = G \times \cos(2\pi f t) \times n' \times \psi \qquad (2)$$

where:
$G \times \cos(2\pi f t)$ is the harmonic load due to a single pedestrian,
G is the considered weight of a person,
f is the natural frequency under consideration,
n' is the equivalent number of persons (pedestrians or joggers) on the loaded surface S,
S is the loaded surface (according to some approaches [13] it depends on the shape of the normal mode under consideration, according to others [14] the whole 'loadable' surface should be considered),
ψ is the reduction coefficient to take into account the probability that the footfall frequency approaches the natural frequency under consideration. This coefficient is different for each of the load models given below.

The load value depends on the crowd density and on the walking behaviour. When the density is less than 1 P/m² people are not interacting. If the density increases free walking is no longer possible an the interaction of walking people should be considered. Two groups of load models in the guideline are presented here additionally the guideline provides load models for single pedestrians and for joggers.

1.5.1.2 Load model for TC1 to TC3 (density d < 1.0 P/m²)

A uniformly distributed harmonic load $p(t)$ represents the equivalent pedestrian stream for further calculations. The load model for pedestrian groups takes into account a free movement of the pedestrians. And so the synchronization among the group members is equal to a low density stream

Static force of a single pedestrian G, equivalent number of pedestrians n' (95% percentile) and reduction coefficient ψ are given in Table 1.5 [14].

1.5.1.3 Load model for TC 4 and TC 5 (density d ≥ 1.0 P/m²)

In the case of a heavy congestion, walking is obstructed: the moving forward is slow and the synchronisation increases. Beyond the upper limit values for the dense stream (up to 1.2 or 1.5 or even 2 [15] Persons/m²), walking becomes impossible therefore significantly reducing dynamic effects. When a stream becomes dense, the correlation between pedestrians increases, but the dynamic load tends to decrease.

The amplitude of the dynamic force of a single pedestrian G, equivalent number of pedestrians n' (95 % percentile) and reduction coefficient ψ are given in Table 1.6 [14].

Table 1.5 Parameters for Load models of TC1 to TC3.

G [N]			
Vertical	Longitudinal	Lateral	$n'\,[1/m^2]$
280	140	35	$\dfrac{10.8\sqrt{\xi \times n}}{S}$

Reduction coefficient (ψ)

Vertical and longitudinal	Lateral

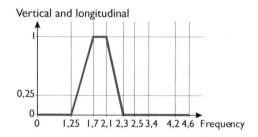

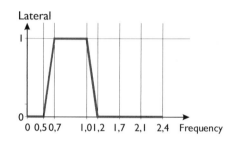

where:
ξ is the structural damping and
n is the number of the pedestrians on the loaded surface S ($n = S \times$ density).

Table 1.6 Parameters for Load model of TC4 and TC5.

G [N]			
Vertical	Longitudinal	Lateral	$n'\,[1/m^2]$
280	140	35	$1.0 \times 1.85\sqrt{n}$

Reduction coefficient ψ

Vertical and longitudinal	Lateral

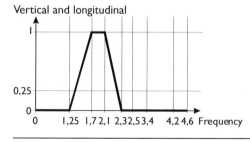

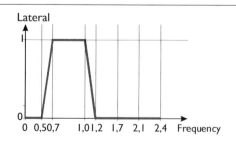

where:
n is the number of the pedestrians on the loaded surface S ($n = S \times$ density).

1.5.2   Evaluation methods

1.5.2.1   Spectral evaluation

The spectral evaluation method [16] is the result of an extensive study of the system response to various pedestrian streams. The aim of the development of a spectral design model was to find in a simple way a descritpion of the stochastic loading and system response that gives design values with a specific confidence level. The general design procedure is adopted from wind engineering where it is used to verify the effect of gusts on sway systems.

As design value the system response "maximum peak acceleration" was chosen. In the design check this acceleration is compared with the tolerable acceleration according to the comfort level to be verified.

This maximum acceleration is defined by the product of a peak factor k_a and a derivation of acceleration σ_a.

Both factors have been derived from Monte Carlo simulations which are based on numerical time step simulations of various pedestrian streams on various bridges geometries.

The bases of the variance of acceleration are the stochastic loads. To determine these loads bridges with spans in the range of 20 m to 200 m and a varying width of 3 m and 5 m with four different stream densities (0.2 Pers/m², 0.5 Pers/m², 1.0 Pers/m² and 1.5 Pers/m²) have been investigated. For each bridge type and stream density 5000 different pedestrian streams have been simulated in time step calculations where each pedestrian had the following properties which are taken randomly from the specific statistical distribution:

- Persons weight (mean = 74,4 kg; standard deviation = 13 kg),
- Step frequency (mean value and standard deviation depend on stream density),
- Factor for lateral foot fall forces (mean = 0,0378, standard deviation = 0,0144),
- Start position (randomly), and
- Moment of first step (randomly).

The result is the characteristic acceleration, which is the 95% fractile of the maximum acceleration. For different pedestrian densities it can be determined according to the following formulas and tables while it is assumed that

- the mean step frequency of the pedestrian stream coincides with the considered natural frequency of the bridge,
- the mass of the bridge is uniformly distributed,
- the mode shapes are sinusoidal,
- no modal coupling exists,
- the structural behaviour is linear-elastic.
- for vertical vibrations $f_{s,5\%,slow} = 1,25 \text{ Hz} \leq f_i \leq 2,3 \text{ Hz} = f_{s,95\%,fast}$
- for lateral vibrations $0,5 \text{ Hz} \leq f_i \leq 1,2 \text{ Hz} = f_{s,95\%,fast/2}$

Note: 0,5 Hz is chosen because that natural frequency was excited during the inauguration of the Millennium Bridge.

The following empirical expression for the determination of the variance of the response is recommended:

$$a_{max} = a_{max,95\%} = K_{a,95\%} \frac{d \cdot l \cdot b}{M_i} \sqrt{C \cdot K_f^2 K_1 \xi^{K_2}} \tag{3}$$

where:

K_1 : constant $K_1 = a_1 f_i^2 + a_2 f_i + a_3$

K_2 : constant $K_2 = b_1 f_i^2 + b_2 f_i + b_3$

a_1, a_2, a_3 : constants, see Table 1.7 for vertical and Table 1.8 for lateral vibrations

b_1, b_2, b_3 : constants, see Table 1.7 for vertical and Table 1.8 for lateral vibrations

$n = d \cdot l \cdot b$: number of persons on the bridge, d: pedestrian density, l: bridge length, b: bridge width

K_F : constant

σ_F^2 : variance of the loading (pedestrian induced forces)

f_i : considered natural frequency that coincides with the mean step frequency of the pedestrian stream

M_t : modal mass of the considered mode i

ξ : damping ratio

C : constant describing the maximum of the load spectrum

σ_a^2 : variance of the acceleration response

$K_{a,94\%}$: peak factor to transform the standard deviation of the response a to the characteristic value $a_{max,95\%}$

The constants a_1 to a_3, b_1 to b_3, C, k_F and $k_{a,95\%}$ can be found in Table 1.7 for vertical accelerations and in Table 1.8 for lateral accelerations.

1.5.3   SDOF evaluation

The dynamic behaviour of a structure may be evaluated by modal analysis. Thereby, an arbitrary oscillation of the structure is described by a combination of n different harmonic oscillations with different frequencies for each mode shape i. In doing so,

Table 1.7 Constants for vertical accelerations.

d [P/m²]	k_F	C	a_1	a_2	a_3	b_1	b_2	b_3	$k_{a,95\%}$
≤0.5	$1.2 \cdot 10^{-2}$	2.95	−0.07	0.6	0.075	0.003	−0.04	−1	3.92
1.0	$7 \cdot 10^{-3}$	3.7	−0.07	0.56	0.084	0.004	−0.045	−1	3.80
1.5	$3.335 \cdot 10^{-3}$	5.1	−0.08	0.5	0.085	0.005	−0.06	−1.005	3.74

Table 1.8 Constants for lateral accelerations.

d [P/m²]	k_F	C	a_1	a_2	a_3	b_1	b_2	b_3	$k_{a,95\%}$
≤0.5		6.8	−0.08	0.5	0.085	0.005	−0.06	−1.005	3.77
1.0	$2.85 \cdot 10^{-4}$	7.9	−0.08	0.44	0.096	0.007	−0.071	−1	3.73
1.5		12.6	−0.07	0.31	0.12	0.009	−0.094	−1.02	3.63

the structure is divided into n different spring mass oscillators, each with a single degree of freedom (SDOF). Each SDOF oscillator has a modal mass M_i, a modal stiffness K_i and a modal load P_i. The equivalent spring mass system is found with the method of generalization, Fig. 1.4.

The basic idea is to use one single equivalent SDOF system for each natural frequency i of the footbridge in the critical range and to calculate the associated maximum acceleration for a dynamic loading. The maximum acceleration a_{max} in the resonant for the SDOF is calculated by:

$$a_{max} = \frac{P_i \cdot \pi}{M_i \cdot \delta} = \frac{P_i}{M_i \cdot 2\xi} \tag{4}$$

As a simple example, a single span beam, Figure 1.5, is considered. This beam has a distributed mass µ [kg/m] a stiffness k and a length l. The uniform load $p(x)$ sin (ωt) is distributed over the total length. The mode shapes $\phi(x)$ of the bending modes are assumed to be a half sin distribution $\phi(x) = sin(m^*x/l^*\pi)$ whereas m is the number of half waves which is here equal to the number of mode shape i. The load oscillates with sin (ωt).

The generalized mass M_i and generalized load P_i sin(ωt) of the SDOF system are calculated for a single span beam with a harmonic uniform load p·sin (ωt) according to Table 1.9. The generalized load for a single load P_{mov} sin (ωt), moving across the simple beam is also given in Table 1.9. This excitation is limited by the tuning time which is defined as the time for the moving load to cross one belly of the mode shape.

The 2nd mode shape $i = 2$ of a single span beam has two half waves. Loading the entire length, the generalized load will be calculated to a value of $P_i = 0$, when half of the uniformly distributed load is acting against the displacements of one belly and the other half is acting within the direction of displacements. The generalized load according to Table 1.9 is based on the assumption that only one belly of the mode

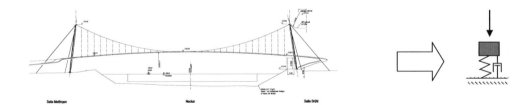

Figure 1.4 One equivalent SDOF oscillator.

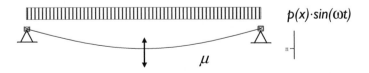

$p(x)\cdot sin(\omega t)$

Figure 1.5 Simple beam with harmonic mode shape ϕ (x), i = 1.

Table 1.9 Generalized mass and generalized load.

Mode shape	Generalized mass M_i	Generalized load P_i for uniformly load $p(x)$ P_i	Generalized P_i load for moving load P_{mov} P_i	Tuning time t_{max}
$i = 1:\ \Phi(x) = \sin\left(\dfrac{x}{l}\pi\right)$	$\dfrac{1}{2}\mu l$	$\dfrac{2}{\pi}p(x)\,l$	$\dfrac{2}{\pi}P_{mov}$	l/v
$i = 2:\ \Phi(x) = \sin\left(\dfrac{2x}{l}\pi\right)$	$\dfrac{1}{2}\mu l$	$\dfrac{1}{\pi}p(x)\,l$	$\dfrac{2}{\pi}P_{mov}$	$l/(2v)$
$i = 3:\ \Phi(x) = \sin\left(\dfrac{3x}{l}\pi\right)$	$\dfrac{1}{2}\mu l$	$\dfrac{2}{3\pi}p(x)\,l$	$\dfrac{2}{\pi}P_{mov}$	$l/(3v)$

where:
P_{mov}	is the moving load in kN		l	is the length of the bridge	
$p(x)$	is the distributed load in kN/m		i	is the mode shape number (half wave)	
μ	is the mass distribution per length		v	is the velocity of the moving load	

shape is loaded, which results in larger oscillations. A more conservative approach to distribute the loading on the whole beam is favored by the SETRA/AFGC guidelines. There it is recommended to take the total length of the beam into account independent from the mode shape. Then, the load is always acting in the direction of displacements of the bellies and the generalized load P_i for all mode shapes is the same as for the first bending mode ($i = 1$).

1.5.4 Finite element evaluation

Nowadays, even conceptual design of footbridges takes advantage of using the finite element method (FEM). Hence, preliminary dynamic calculations may easily be performed without additional effort. A simple approach to perform the static and dynamic calculation is by modeling the bridge deck by beam elements and the cable with cable elements, spring or truss elements in a three dimensional FEM model. A rough overview of the natural frequencies and the appropriate mode shapes is obtained and possible problems in dynamic behavior can be identified.

The non load bearing parts such as furniture and railings are considered as additional masses as exactly as possible. A more refined model may take advantage of various types of finite elements such as plate, shell, beam, cable or truss elements. The more complex the static system and the higher the mode shape, the more finite elements are required. The model should always allow for possibly vertical, horizontal, and torsional mode shapes.

To get reliable results for natural frequencies, it is absolutely necessary that stiffness and mass distribution are modeled in a realistic way. All dead load, superimposed dead load and pre-stressing of cables have to be considered for the calculation of natural frequencies. A lumped mass approach, in which rotational masses are neglected, is in many cases sufficient. The modal mass regarding to each mode shape should be available, when verification of comfort is done with simple approaches by hand calculation.

Many parameters such as properties of materials, complexity of the structure, the type of deck surfacing and furniture, boundary conditions and railings may cause discrepancies in natural frequencies between the results of computer calculations and the measured data of the real structure.

A numerical modeling should be as realistic as possible with regard to bearing conditions and their foundation stiffness. For the modeling of abutments and foundations, dynamic soil stiffness should be used. Otherwise the obtained results will be very inaccurate.

1.6   IMPROVING THE STRUCTURE

If the dynamic response of the structure under a specified traffic load does not fulfill the comfort requirements as specified it is necessary to improve the structure. In general there are three means to do this:

- Modification of model mass
- Modification of natural frequency
- Installation of additional damping devices.

Modal mass and natural frequency can be modified at the design stage only. For an already constructed bridge, the simplest approach is based on the increase of the structural damping, which can be achieved either by implementation of control devices, or by actuation on non-structural finishings, like the hand-rail and surfacing.

1.6.1   Modification of model mass

For very light footbridges e.g., the use of heavy concrete deck slabs can improve dynamic response to pedestrian loads, as consequence of the increased modal mass.

On the basis of the spectral design model, Butz[16] developed (see 5.2.1) also an empirical expression for the determination of a required modal mass for a given pedestrian traffic to ensure a required comfort a_{limit} that is valid for *mean pace frequency $f_{s,m}$ = Natural frequency i of the bridge f_i*:

$$M_i \geq \frac{\sqrt{n}(K_1 \xi^{K_2} + 1.65 K_3 \xi^{K_4})}{a_{limit}} \tag{5}$$

where M_i modal mass for considered mode i
        n number of pedestrians on the bridge
        ξ damping coefficient
        K_1 to K_4 constants (see *Table 1.10* and *Table 1.11*)

1.6.2   Modification of natural frequency

The classical proposal of modification of structural properties in order to avoid natural frequencies in the critical range for vertical and lateral vibrations does not meet

Table 1.10 Constants for required vertical modal mass (vertical bending and torsion modes).

d [P/m²]	k_1	k_2	k_3	k_4
≤0,5	0,7603		0,050	
1,0	0,570	0,468	0,040	0,675
1,5	0,400		0,035	

Table 1.11 Constants for the required lateral modal mass (horizontal bending modes).

d [P/m²]	k_1	k_2	k_3	k_4
≤0,5				
1,0	0,1205	0,45	0,012	0,6405
1,5				

at the current state-of-art the goal of bridge designers to build light and graceful structures. In effect, given the proportionality of natural frequencies to the square root of the ratio between stiffness and mass, it is understandable that considerable modifications are required in the stiffness in order to attain a slight increase of natural frequency. However, it is of interest to consider during the design stage several simple strategies that can improve the dynamic behaviour and that are normally associated with an increase of critical natural frequencies. These comprehend, for example, the replacement of a reinforced concrete deck slab formed by non-continuous panels by a continuous slab, or the inclusion of the handrail as a structural element, participating to the overall deck stiffness.

Other more complex measures can be of interest, like the addition of a stabilizing cable system. For vertical vibrations, alternatives are the increase of depth of steel box girders, the increase of the thickness of the lower flange of composite girders, or the increase of depth of truss girders. For lateral vibrations, the most efficient measure is to increase the deck width. In cable structures, the positioning of the cables laterally to the deck increases the lateral stiffness. In cable-stayed bridges, a better torsional behaviour can be attained by anchoring of the cables at the central plane of the bridge on an A-shape pylon, rather than anchoring them at parallel independent pylons.

1.6.3   Installation of additional damping devices

The increase of structural damping is another possible measure to reduce dynamic effects of pedestrian movements on footbridges. This increase can be achieved either by actuation on particular elements within the structure, or by implementation of external control devices.

The use of external damping devices for absorbing excessive structural vibrations can be an effective solution in terms of reliability and cost. These devices can be based on active, semi-active or passive control techniques. Considering aspects like cost, maintenance requirements and practical experience, the usual option is for passive devices, which comprehend viscous dampers, tuned mass dampers (TMDs), pendulum dampers, tuned liquid column dampers (TLCDs) or tuned liquid (TLDs). The most popular of these are viscous dampers and TMDs.

1.7   CONCLUSIONS

This paper presents a procedure for the design of footbridges with regard to comfort criteria. A relevant and new aspect in the design procedure is the design first step in which comfort criteria for different traffic situations are fixed in communication with the client. It allows light and graceful structures by fixing realistic and adequate requirements.

Additionally different methods for the determination of the relevant accelerations are presented. These design methods cover simple evaluation methods e.g. for preliminary design studies and more complex methods for the detailed layout.

The paper finishes with the presentation of different methods of improvement of the structural properties such that comfort requirements are fulfilled.

Application examples can be found in the SYNPEX final report [1] and will be published in the scope of the HiVoSS project mentioned below.

The guideline presented here will also be published together with a guideline for vibration design for floors within the scope of the RFCS project HiVoSS in the languages English, French, Portuguese, Dutch and German. Also seminars on the vibration design of footbridges and floors have been performed in different countries. Further information can be found in the internet searching for "HiVoSS".

ACKNOWLEDGEMENTS

The design guidance presented here was elaborated within the research project "Advanced Load Models for Synchronous Pedestrian Excitation and Optimised Design Guidelines for Steel Footbridges (SYNPEX)" which was performed with the grant of the Research Programme of the Research Fund for Coal and Steel (RFCS) of the European Community (Project No RFS-CR-03019).

Special thanks apply for the project partners Christiane Butz (MAURER & SÖHNE), Elsa Caetano (FEUP), Alvaro Cunha (FEUP), Arndt Goldack (SCHLAICH BERGERMANN & PARTNER), Andreas Keil (SCHLAICH BERGERMANN & PARTNER) and Mladen Lukic (CTICM) who have a large portion of the results presented in this paper.

REFERENCES

[1] Butz, CH.; Heinemeyer, CH.; Goldack, A.; Keil, A.; Lukic, M.; Caetano, E.; Cunha, A.: *Advanced Load Models for Synchronous Pedestrian Excitation and Optimised Design Guidelines for Steel Footbridges (SYNPEX)*; RFCS-Research Project RFS-CR-03019, will soon be found in http://bookshop.europa.eu/

[2] BRITISH STANDARDS INSTITUTION: BS5400, Part 2, *Appendix C: Vibration Serviceability Requirements for Foot and Cycle Track Bridges*, Great Britain, 1978.

[3] DEUTSCHES INSTITUT FUER NORMUNG: *DIN-Fachbericht 102*, Betonbrücken, 2003.

[4] EUROPEAN COMMITTEE FOR STANDARDIZATION CEN: ENV 1995-2, *Eurocode 5 – Design of timber structures – bridges*, 1997.

[5] FIB: *Guidelines for the design of footbridges*, fib bulletin 32, November 2005.

[6] EUROPEAN COMMITTEE FOR STANDARDIZATION CEN: prEN1991-2:2002, *Eurocode 1 – Actions on structures*, Part 2: Traffic loads on bridges, 2002.

[7] EUROPEAN COMMITTEE FOR STANDARDIZATION CEN: prEN1995-2, *Eurocode 5 – Design of timber structures*. Part 2: Bridges, 2003.

[8] EUROPEAN COMMITTEE FOR STANDARDIZATION CEN: prEN1998-2:2003, *Eurocode 8 – Design of structures for earthquake resistance*, Part 2: Bridges, 2003.

[9] Fitzpatrick, T. et al.: *Linking London: The Millennium Bridge*, The Royal Academy of Engineering, London, 2001, ISBN 1 871634 997.

[10] Charles, P.; Bui, V.: *Transversal dynamic actions of pedestrians & Synchronisation*, Proceedings of Footbridge 2005 – 2nd International Conference, Venice 2005.

[11] SETRA/AFGC: *Passerelles piétonnes – Evaluation du comportement vibratoire sous l'action des piétons (Footbridges – Assessment of dynamic behaviour under the action of pedestrians)*, Guidelines, Sétra, March 2006.

[12] Bachmann, H.; W. Ammann, *Vibrations in Structures Induced by Man and Machines*. IABSE Structural Engineering Documents, 1987. No. 3e.

[13] SETRA/AFGC: Comportement *Dynamique des Passerelles Piétonnes (Dynamic behaviour of footbridges)*, Guide (Draft), 15 December 2004.

[14] SETRA/AFGC: Comportement Dynamique *des Passerelles Piétonnes (Dynamic behaviour of footbridges)*, Guide (Final draft), January 2006.

[15] Fujino Y. et al.: Synchronisation of human walking observed during lateral vibration of a congested pedestrian bridge, *Earthquake Engineering and Structural Dynamics*, Vol.22, pp. 741–758, 1993.

[16] Butz, Ch.: *Beitrag zur Berechnung fußgängerinduzierter Brückenschwingungen*, Schriftenreihe des Lehrstuhls für Stahlbau und Leichtmetallbau der RWTH Aachen Heft 60, 2006, ISBN 3-8322-5699-7.

Chapter 2

Application of French guidelines in design

Pascal Charles
Setra, Bagneux, France

SUMMARY

This paper presents an example of the application of the French guidelines concerning the dynamic behaviour of footbridge, to a real footbridge on the river Seine near Paris, which spans are 50 m–100 m–50 m long. The French methodology is applied from the beginning to the end, including the choices of the owner (traffic and level of comfort), the dynamic calculations (modes shapes, eigenfrequencies, accelerations due to dynamic loads), and the consequences to the comfort and the possibilities to improve the dynamic behaviour of the footbridge.

This footbridge presents two dangerous modes for the comfort, a lateral one that is sensitive to the lock-in effect (the natural frequency is 1.06 Hz) and a vertical one that is sensitive to the classical effects of the vertical dynamic force of the steps of the pedestrians. The lateral vibration problem is solved by connecting the footbridge to the concrete bridge next to it. The vertical vibration problem is solved by increasing the mass of the footbridge, or by adding Tuned Mass Damper.

Keywords: Footbridge; dynamic; vibration.

2.1 INTRODUCTION

In 2001, The Sétra and the French civil engineering association have assigned a task group for the appraisal of the dynamic behaviour of footbridges. A research program was initiated by the French Road Directory to organize field tests. Two structures were investigated to identify their vibration behaviour under pedestrian loading: the Solferino footbridge and a full seal model representing a swaying footbridge.

Guidelines have been drawn up on the basis of current technical and scientific knowledge gained in France and out of the country. These guidelines synthesize the dynamic problems of footbridges and present a dynamic analysis methodology with the purpose to provide information to assist designers and engineers to better understand the phenomena and criteria to help them with the design. These guidelines were presented during the last "footbridge" conference in Venice in 2005 and published in 2006[1]. They are now applied in France for two years without any other event, or difficulties to design footbridges.

These guidelines deal with normal use towards comfort criteria, and vandalism cases towards structural integrity, the case of exceptional events (marathon, dance, military march-past) is not considered.

The dynamic analyses methodology depends on footbridge category reliable to the traffic level; which means that urban footbridges are not considered like rural ones. In fact, the analysis proposed is a risk analysis. The owner must do the choice of the comfort criteria, which has influence on the design. For a maximum comfort no vibration is allowed, for a minimum comfort, moderate and controlled vibrations are allowed.

This paper presents an example of a footbridge which has been verified with these rules. The French guidelines are reminded while applied.

2.2 PRESENTATION OF THE EXAMPLE

The example is a footbridge situated on the river Seine, near Paris, in a town named "Bezon". The footbridge is situated on both lateral sides of an existing bridge (Figure 2.1, Figure 2.2) which is not enough wide to accept road traffic, railway traffic and pedestrians. So two footbridges which are 4 meters wide, and with the same spans (3 spans: 50 meters–100 meters–50 meters) than the existing concrete bridge are planned. The structure presented here (that was not the structure finally chosen for the building) is a continuous steel box girder that is 1 m 50 high. The intermediate existing piers are wide enough to support the footbridge. Just at the level of these piers, the footbridge is reinforced with a trellis. On the deck of the footbridge, there is a connected concrete slab that is 10 cm thick.

2.3 APPLICATION OF THE FRENCH METHODOLOGY

The French methodology is summarized with the chart presented in Figure 2.3.

2.3.1 Step 1: Setting of the footbridge class

The footbridge class allows placing the traffic level expected.

Class IV: low usage (footbridge), built in a rural environment, linking sparsely populated zones or providing continuity to the pedestrian traffic over an express way or a railway.

Class III: normal usage (footbridge), with important group of pedestrians crossing occasionally the structure, but never on the whole deck area.

(a)

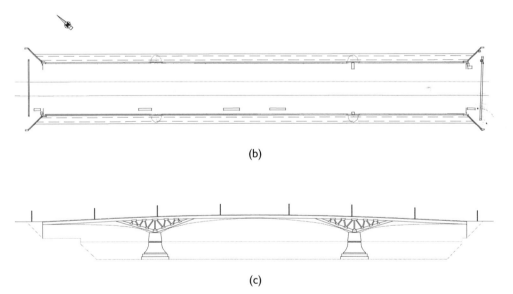

(b)

(c)

Figure 2.1 Existing bridge: the existing piers are wide enough to support the future footbridge which will be built next to the concrete bridge: (a) general view; (b) plan view; (c) lateral view.

Class II: urban (footbridge), linking populated zones, with large traffic occasionally on the whole deck area.

Class I: urban (footbridge) with very large traffic, linking high concentration populated zones (close to a railway or metro station) or with crowds crossing frequently the structure.

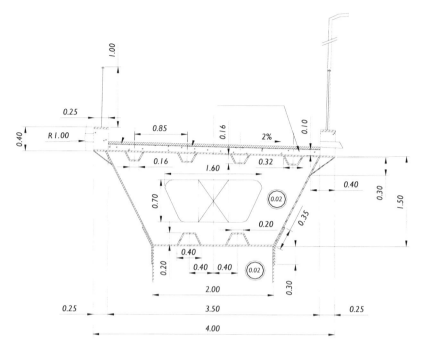

Figure 2.2 Cross section of existing bridge.

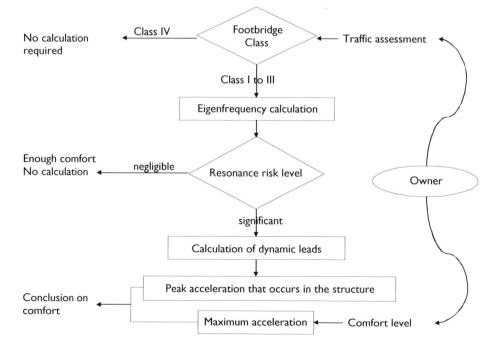

Figure 2.3 Methodology organization chart.

It is up to the owner to decide to set the footbridge class based on the above and according the potential traffic development. His choice could be influenced by other criteria. For instance, an "outclass" could be adopted to increase the prevention level against vibrations, in case of strong media stakes. On the contrary, a "subclass" could be adopted to reduce the construction costs, or to allow a greater freedom in design; of course, the risk of "subclass" is limited, in case of loading exceeding the common values of traffic and magnitude, to discomfort feeling for a few persons.

The footbridge in class IV doesn't require calculation towards dynamic behaviour. For very light footbridges it seems careful to adopt at least the class III, for a minimum risk control. However, very light footbridge can have high acceleration without resonant phenomena.

For the "Bezon" footbridge, the classes I, II and III are studied in order to determine the consequences of the choices of the owner.

2.3.2 Step 2: Choice of the comfort level

The owner sets the comfort level for the footbridge:

Maximum comfort: accelerations undergone by the structure are un-significant for users.

Mean comfort: accelerations undergone by the structure are simply felt by users.

Minimum comfort: in case of infrequent loading, accelerations undergone by the structure are felt by users, but not necessary intolerable.

Tables 2.1 and 2.2 define, respectively for vertical and horizontal accelerations, four value grades identified as 1, 2, 3 and 4. The first three values correspond to, in ascending order, the maximum, mean, and minimum comfort. The fourth grade corresponds to the uncomfortable acceleration levels that are not tolerated.

For the "Bezon" footbridge, a "mean" comfort is decided. The maximum confort is chosen in very specific case (school, hospital near the footbridge), and the minimum confort when specific measures are taken in order to warn people who cross the footbridge.

Table 2.1 Acceleration grading for vertical vibrations.

Acceleration (m/s²)	0	0.5	1.0	2.5
Grade 1	Max			
Grade 2		Mean		
Grade 3			Min	
Grade 4				

Table 2.2 Acceleration grading for horizontal vibrations. The acceleration is limited in any case to 0.1 m/s² to avoid lock-in effect.

Acceleration (m/s²)	0	0.1	0.15	0.3	0.8
Grade 1	Max				
Grade 2			Mean		
Grade 3				Min	
Grade 4					

2.3.3   Step 3: Frequency setting – and calculation needed or not with dynamic loads

For footbridges in class I to III, it is necessary to determine the vibration eigenfrequency of the structure. These frequencies affect the three vibrations: vertical, transversal horizontal and longitudinal horizontal. They are calculated for two values of mass: empty footbridge, footbridge loaded with one pedestrian per square meter on the whole surface deck (70 kg/m²). For the "Bezon" footbridge, which is a very heavy footbridge, the mass of the pedestrian is unsignificant, so only one calculation is performed.

Frequency grades allow estimating the risk of resonance induced by pedestrian traffic and fixing the dynamic loads calculation to verify the comfort criteria. Frequencies are classified in four grades in both directions (vertical, horizontal) (see Tables 2.3 and 2.4). The highest grade, ("1"), is corresponding to the maximum risk to be at resonance.

Table 2.3 Grading frequencies for vertical and longitudinal vibrations.

Frequency (Hz)	0	1.0	1.7	2.1	2.6	5
Grade 1						
Grade 2						
Grade 3						
Grade 4						

Table 2.4 Grading frequencies for transversal horizontal vibrations.

Frequency (Hz)	0	0.3	0.5	1.1	1.3	2.5
Grade 1						
Grade 2						
Grade 3						
Grade 4						

Table 2.5 Load cases for accelerations control.

Traffic	Class	Grade where is the eigenfrequency		
		1	2	3
Not Very dense	III	Case 1	None	None
Dense	II	Case 1	Case 1	Case 3
Very dense	I	Case 2	Case 2	Case 3

Case 1: Not very dense and dense crowd;
Case 2: Very dense crowd;
Case 3: Additional crowd (2nd harmonic).

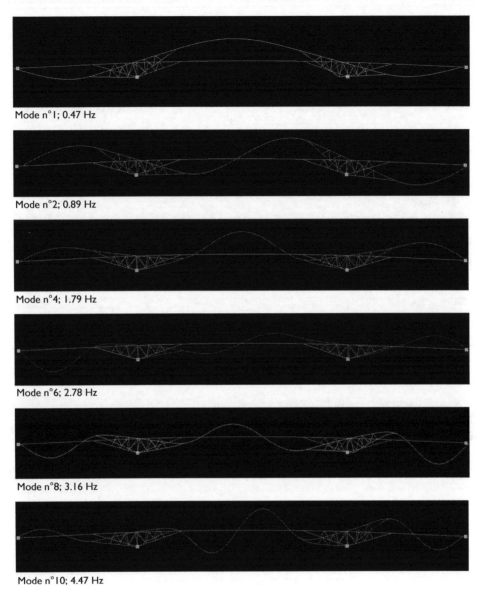

Mode n°1; 0.47 Hz

Mode n°2; 0.89 Hz

Mode n°4; 1.79 Hz

Mode n°6; 2.78 Hz

Mode n°8; 3.16 Hz

Mode n°10; 4.47 Hz

Figure 2.4 Vertical modes, lateral view.

The following grades are corresponding, in descending order, to mean risk ("2"), minimum risk ("3"), and negligible risk ("4").

Definition of dynamic calculations

Depending on the footbridge class and where the eigenfrequency is graded, it could be necessary to carry out dynamic calculations with different load cases. Table 2.5

systematises those cases as function of the eigenfrequency grade and the footbridge class.

For the Bezon footbridge the first 10 frequencies are systematised in Table 2.6, while the mode shape configurations are represented in Figures 2.4 and 2.5.

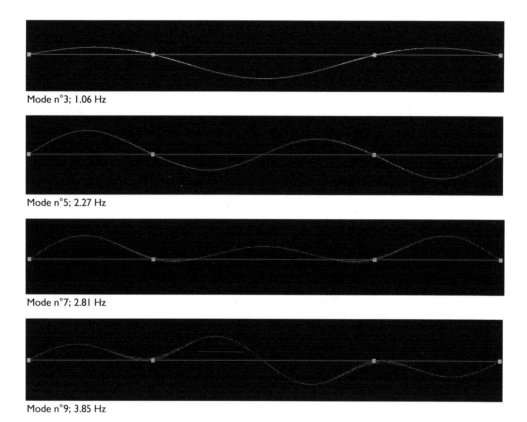

Mode n°3; 1.06 Hz

Mode n°5; 2.27 Hz

Mode n°7; 2.81 Hz

Mode n°9; 3.85 Hz

Figure 2.5 Transversal modes, top view.

Table 2.6 Frequencies and characteristics of first 10 modes.

Mode N°	Pulsation (Rad/s)	Period (s)	Frequency (Hz)	Direction
1	2.98	2.107	0.47	Vertical
2	5.58	1.127	0.89	Vertical
3	6.65	0.944	1.06	Transversal
4	11.22	0.56	1.79	Vertical
5	14.26	0.441	2.27	Transversal
6	17.48	0.359	2.78	Vertical
7	17.66	0.356	2.81	Transversal
8	19.88	0.316	3.16	Vertical
9	24.17	0.26	3.85	Transversal
10	28.06	0.224	4.47	Vertical

The mode 1 and 2 are outside the dangerous range of frequencies (grade 4).

The third mode is the first transversal mode, with a frequency of 1.06 Hz. It is very dangerous for the comfort (grade 1), because a lock-in effect could occur.

The fourth mode is a vertical mode at 1.79 Hz. It is also a dangerous mode for the comfort because it is in grade 1.

The fifth mode is a transversal mode which is at the limit of the grade 3 range (2.27 Hz just before the 2.5 Hz limit). It can only be excited by the second harmonic of pedestrian step.

The modes 6, 8 and 10 are vertical modes in grade 3. They are only sensitive to the second harmonic of the pedestrian step.

The modes 7 and 9 are outside the dangerous range. They are grade 4 modes, and are not to be taken into account.

The other modes have frequencies that are above 5 Hz, and are consequently not to be taken into account.

Finally, only the modes 3, 4, 5, 6, 8 and 10 are to be taken into account. The two first ones are the most dangerous.

If the Bezon footbridge is a class I footbridge, all these modes need calculations with a very dense crowd.

If the Bezon footbridge is a class II footbridge, all these modes need calculations with a dense crowd.

If the Bezon footbridge is a class III footbridge, only the two first modes need calculations with a not very dense crowd.

2.3.4 Step 4: Calculation with dynamic loads

Dynamic loads defined below were developed to represent, in a simple and practical way, pedestrians' effects on the structure. They are constructed for each eigenmode related to a frequency grade with risk to be at resonance.

Case 1: not very dense (class III) and dense crowd (class II), first harmonic.

This case must be considered for footbridges in class III (not very dense crowd) and II (dense crowd). Crowd density, d, depends on the footbridge class and is defined in Table 2.7.

The crowd is distributed on the whole surface of the footbridge S. The number of pedestrians on the footbridge is: $N = S \times d$.

The number of equivalent pedestrians, which means the number of pedestrians, being at the same frequency and phase, producing the same effects those random pedestrians at the same frequency and phase is: $10.8 \times (\xi \times N)^{1/2}$

For the Bezon footbridge, the surface is $S = 4\,m \times 200\,m = 800\,m^2$. The total number of pedestrians, and the equivalent number of pedestrians (with a 0.5% damping ratio) are then defined in Table 2.8.

Table 2.7 Crowd density according to footbridge class.

Class	Crowd density "d"
III	0.5 pedestrian/m²
II	0.8 pedestrian/m²

Table 2.8 Equivalent number of pedestrians for dynamic calculations.

Class of the footbridge	Number of pedestrians N	Equivalent number N_{eq}
II	= 800 × 0.8 = 640	10.8 (640 × 0.005)$^{1/2}$ = 19.3
III	= 800 × 0.5 = 400	10.8 (400 × 0.005)$^{1/2}$ = 15.3

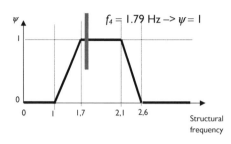
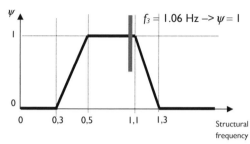

Figure 2.6 Coefficient ψ in the case of walking, for vertical and longitudinal vibrations on the left, and for lateral vibrations on the right.

Table 2.9 Load per length unit associated with random crowd.

Direction	Load per length unit
Vertical (v)	(280 N) × cos($2\pi f_v t$) × 10.8 × (ξN)$^{1/2}$ × ψ/L
Longitudinal (l)	(140 N) × cos($2\pi f_l t$) × 10.8 × (ξN)$^{1/2}$ × ψ/L
Transversal (t)	(35 N) × cos($2\pi f_t t$) × 10.8 × (ξN)$^{1/2}$ × ψ/L

Table 2.10 Equivalent number of pedestrians and amplitude of load for random crowd.

Case	Direction	N_{eq}	Force per length unit
Mode 3, class II	Transversal	19.3	3.38 N/m
Mode 3, class III	Transversal	15.3	2.67 N/m
Mode 4, class II	Vertical	19.3	27.05 N/m
Mode 4, class III	Vertical	15.3	21.38 N/m

The load is modified by a coefficient ψ defined according to the representation of Figure 2.6 conveying the fact that footbridge resonance is less likely when the frequency (vertical or horizontal) is away from the grade 1. The coefficient ψ ranges from 1 to 0 in grade 2. This coefficient becomes null when the footbridge frequency is in grade 3 or 4. However, in this case we'll have to look at the second harmonic of the pedestrians walk.

Table 2.10 summarizes the load per unit of length to apply for each vibration direction, for a random crowd. ξ is the critical damping ratio, and N is the number of pedestrians on the footbridge ($d \times S$).

Loads must be applied on the whole surface of the footbridge deck, and the vibration amplitude sign must be chosen in any point to produce the maximum effect: the

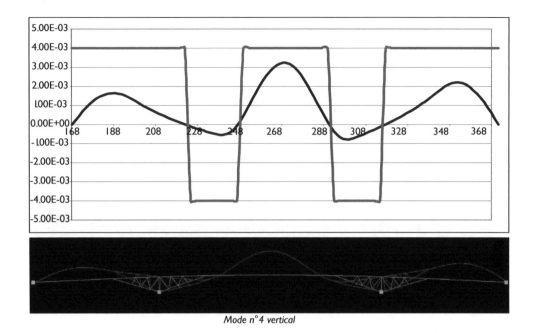

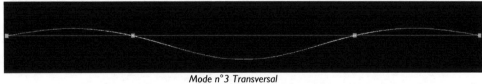

Mode n°4 vertical

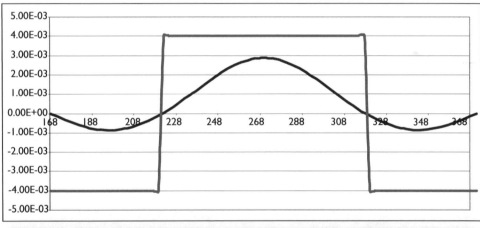

Mode n°3 Transversal

Figure 2.7 Mode shape and shape of distributed load for a vertical and a transversal mode.

loading direction must be in the same way that the mode shape direction, and reverse each time the mode shape is in opposite direction, for instance after a node.

Figure 2.7 presents the mode shape of the deck alone, and the shape of the load that is applied on the footbridge for the mode 3 and 4.

Case 2: very dense crowd (class I) , first harmonic.

This case must be taken into account for a footbridge in class I. Crowd density, d, is 1 pedestrian/m². The crowd is distributed on the whole surface of the footbridge S. We consider pedestrians at the same frequency and presenting random phases. In this case the number of pedestrians at the same phase, equivalent to the number of pedestrians presenting random phases, is $1.85\sqrt{N}$, which gives 52.3 pedestrians for the "Bezon" footbridge. Tables 2.11 and 2.12 systematise the load characteristics and equivalent numbers of pedestrians for this very dense crowd case.

Case 3: distributed crowd 2nd harmonic effect.

This case is similar to cases 1 and 2, but considering the second harmonic of pedestrians walking, about double of the first harmonic frequency. It must be taken into account for a footbridge in class II and I. Crowd density, d, is 0.8 for class II, and 1.0 for class I. The crowd is distributed on the whole surface of the footbridge S.

Table 2.11 Load per length unit associated with very dense crowd.

Direction	Load per length unit
Vertical (v)	$(280 \, N) \times \cos(2\pi f_v t) \times 1.85 \times (N)^{1/2} \times \psi/L$
Longitudinal (l)	$(140 \, N) \times \cos(2\pi f_l t) \times 1.85 \times (N)^{1/2} \times \psi/L$
Transversal (t)	$(35 \, N) \times \cos(2\pi f_t t) \times 1.85 \times (N)^{1/2} \times \psi/L$

Table 2.12 Equivalent number of pedestrians and amplitude of load for very dense crowd.

Case	Direction	N_{eq}	Force per length unit
Mode 3, class I	Transversal	52.3	9.16 N/m
Mode 4, class I	Vertical	52.3	73.26 N/m

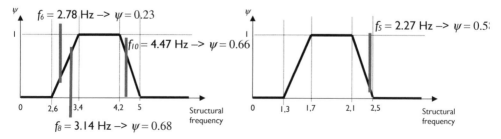

Figure 2.8 Coefficient ψ, for vertical and longitudinal vibrations on the left, and for lateral vibrations on the right.

Table 2.13 Summary of distributed loads for footbridge in classes I and II.

Case		Direction	ψ	N_{eq}	Force per length unit
Mode	5, class I	Transversal	0.58	52.3	1.06 N/m
Mode	5, class II	Transversal	0.58	19.3	0.39 N/m
Mode	6, class I	Vertical	0.23	52.3	4.21 N/m
Mode	6, class II	Vertical	0.23	19.3	1.55 N/m
Mode	8, class I	Vertical	0.68	52.3	12.45 N/m
Mode	8, class II	Vertical	0.68	19.3	4.59 N/m
Mode	10, class I	Vertical	0.66	52.3	12.09 N/m
Mode	10, class II	Vertical	0.66	19.3	4.46 N/m

The pedestrian force is reduced to 70 N in vertical, 35 N in longitudinal, and 7 N in transversal.

The reducing coefficient ψ is given in Figure 2.8. Table 2.13 summarises the amplitudes of distributed loads.

Calculation method

Almost all dynamic structural softwares can perform calculation of the modal caracteristics: modal shape, eigenfrequencies, generalized mass etc... but it is sometimes more difficult to perform dynamic loading, with a very particular load shape. All the loads that are given here above must be applied on the deck of the footbridge, and with the same direction as the modal displacement. The force that is given is the amplitude of the force. Its time-history is so that the structure is excited at its resonance frequency. Then, we know that it is not useful to know the variation of the acceleration. Only the maximum amplitude at resonant peak is needed. This maximum corresponds to the static response in the mode concerned multiplied by the amplification factor.

Then, if v_i are the modal displacements of the finite element number i corresponding to the deck, S_i the surface of the deck concerned by this element, M the generalized mass, p is the amplitude of the dynamic load par m², then the maximum acceleration of the finite element number i is given by:

$$\text{Max}(a(t))_{\text{element } i} = \frac{\sum_{k \in \text{deck}} p\, S_k\, |v_k|}{M} v_i \frac{1}{2\xi}$$

It can be noticed that this formula does not depend on the normalization of the mode.

Then, the maximum acceleration of each point of the deck can be compared to the limit in order to know the level of comfort.

Results

The results presented in Table 2.14 show that the third and fourth mode are able to provoke uncomfort for classes I and II. For class III, comfort is almost assured. In

Table 2.14 Maximum response of footbridge and corresponding comfort limit.

Case	Direction	Frequency	Acceleration	Comfort limit	Status
Mode 3, class I	**Transversal**	**1.06 Hz**	**0.40 m/s²**	**0.1 m/s²**	**Lock-in effect**
Mode 3, class II	**Transversal**	**1.06 Hz**	**0.15 m/s²**	**0.1 m/s²**	**Risk of lock-in effect**
Mode 3, class III	Transversal	1.06 Hz	0.12 m/s²	0.1 m/s²	Nearly OK
Mode 4, class I	**Vertical**	**1.79 Hz**	**3.63 m/s²**	**1 m/s²**	**Very uncomfortable**
Mode 4, class II	**Vertical**	**1.79 Hz**	**1.34 m/s²**	**1 m/s²**	**uncomfortable**
Mode 4, class III	Vertical	1.79 Hz	1.05 m/s²	1 m/s²	Nearly OK
Mode 5, class I	Transversal	2.27 Hz	0.04 m/s²	0.1 m/s²	OK
Mode 5, class II	Transversal	2.27 Hz	0.01 m/s²	0.1 m/s²	OK
Mode 6, class I	Vertical	2.78 Hz	0.20 m/s²	1 m/s²	OK
Mode 6, class II	Vertical	2.78 Hz	0.07 m/s²	1 m/s²	OK
Mode 8, class I	Vertical	3.16 Hz	0.51 m/s²	1 m/s²	OK
Mode 8, class II	Vertical	3.16 Hz	0.19 m/s²	1 m/s²	OK
Mode 10, class I	Vertical	4.47 Hz	0.46 m/s²	1 m/s²	OK
Mode 10, class II	Vertical	4.47 Hz	0.17 m/s²	1 m/s²	OK

class II, comfort is not obtained, but the limit is just exceeded. In practise, accurate measurement of the damping ratio after building should give good results because the slab can participate to a more important damping ratio. But as we do not know if real damping is high enough, damping device must be planned, but installed only in case damping ratio is not high enough.

For class I, modification of the footbridge design or addition of damping devices must be planned.

2.3.5 Step 5: Modification of the footbridge design

Since the previous calculation does not bring verification to a satisfactory conclusion, it's necessary to modify the design.

The modification of the eigenfrequencies is the healthiest way to solve the structural vibration problems. Nevertheless, to modify noticeably the eigenfrequencies, it is necessary to make strong structural reviews to increase the stiffness. For the "Bezon" footbridge, it is in practice easy to treat the first transversal mode because of the proximity of the concrete bridge. A device that prevents dynamic lateral displacement of the footbridge toward the concrete bridge can be installed. If this device is placed at mid span, this mode disappears. The second lateral mode, which remains unchanged because it has a node at midspan, has no vibration problem. This method can not be used for the vertical movement because it is technologically more difficult, and because the risk of the rigidification is that the two first mode which eignefrequencies are 0.47 Hz and 0.89 Hz could go in the dangerous range.

The last solution to regulate the dynamic response of the structure is to use Tuned Mass Dampers. For the class I footbridge, the maximum vertical acceleration is 3.63 m/s² (with a 0.5% damping) whereas the level of comfort is 1 m/s². With a 2% damping ratio, the maximum acceleration goes under 1 m/s². This damping ratio is

Table 2.15 Variation of frequencies due to change of slab thickness.

| Mode N° | Frequency (Hz) | | Direction |
	Thickness of the slab: **0.2 m**	Thickness of the slab: **0.1 m**	
1	0.41	0.47	Vertical
2	0.77	0.89	Vertical
3	**0.91**	**1.06**	**Transversal**
4	**1.54**	**1.79**	**Vertical**
5	1.96	2.27	Transversal
6	2.39	2.78	Vertical
7	2.42	2.81	Transversal
8	2.74	3.16	Vertical
9	3.31	3.85	Transversal
10	3.86	4.47	Vertical

easy to reach with a TMD (that can easily provide damping up to 3 to 4%). Only one mode has to be treated.

Another solution can consist in reinforcing the concrete slab, and use a 0.2 m thick slab instead of 0.1 m. This increase in the mass of the footbridge causes a decrease of the frequencies (see Table 2.15), but also a decrease of the acceleration. As the reinforced concrete damping ratio is higher, this could also provide a higher global damping ratio, but it is difficult to estimate the impact at this level of the studies. The variation of the frequencies are the following:

For the third mode, the acceleration is changed and become 0.11 m/s² instead of 0.15 m/s² for a class II footbridge, which is nearly acceptable. For a class I footbridge, the acceleration remains important: 0.30 m/s² instead of 0.40 m/s² but always superior to the 0.10 m/s² threshold.

For the fourth mode, the decrease of the frequency down to 1.54 Hz instead 1.79 Hz is very interessant, because the ψ factor is then under 1 (0.77). The acceleration is 0.75 m/s² for a class II footbridge (instead of 1.34 m/s²) and 2.04 m/s² instead of 3.63 m/s² for a class I footbridge.

The conclusions for the comfort for other modes are not changed.

The increase of the thickness of the slab is not sufficient for the class I but is sufficient for a class II footbridge. This increase in the mass has an important cost, but is more durable than the TMD.

2.4 CONCLUSION

This paper has given an example of the application of the French recommandations concerning vibrations problems of footbridge. The footbridge that was tested is a real footbridge, and has been tested with 3 different traffic classes.

If the footbridge is a class I footbridge (very dense crowds cross the footbridge), it is compulsary to put TMD to reach a level of damping ratio at about 2%.

If the footbridge is a class II footbridge (dense crowds cross the footbridge), an increase in mass can be sufficient. Even without this increase of mass, accelerations are not too important.

If the footbridge is a class III footbridge, the dynamic behaviour of the footbridge is acceptable.

This example illustrates that the role of the owner of the footbridge who determines both the traffic level and the level of comfort is very important. These rules are used in France since 2006 without any problem since the Solferino footbridge in 2000. On the other hand, the French recommandations are not too strict, and leave possibilities for aesthetic design.

REFERENCES

[1] SETRA/AFGC: *Passerelles piétonnes – Evaluation du comportement vibratoire sous l'action des piétons (Footbridges – Assessment of dynamic behaviour under the action of pedestrians)*, Guidelines, Sétra, March 2006.

Chapter 3

Design for dynamic effects in long span footbridges

Angus low
Arup, London, UK

SUMMARY

Clients are requiring longer spans for footbridges and cycle bridges. The paper follows a number of case histories of recent long-span footbridges. Four dynamic effects are identified which need to be considered by designers and each is discussed. Some of the design criteria are soft, in the sense that they are governed by human issues. The author makes a plea that these issues are not over-specified because the economic feasibility of longer spans may be at risk. There are some general comments on the likely form of future designs.

Keywords: Footbridge, long span, synchronous lateral excitation, pedestrian excitation, vortex excitation.

3.1   INTRODUCTION

There is a growing demand for long-span footbridges and cycle bridges. As highways get wider, and supports within central reserves are seen as less acceptable, so spans of 50 m and more are becoming common. Beyond this there is a growing recognition of the benefits of foot and cycle bridges as generators of more vibrant cities and towns. Many urban areas are split by a major feature – a river or a harbour. A new cycle bridge will naturally generate new routes for many kilometres on each side, using existing streets. These routes can be quiet and pleasant, away from the noise of the through motor traffic. More people will discover the joys of walking and cycling. The frontages will have new value because the passing trade is free to stop and spend money. As a cyclist the author hopes that more local authorities will recognise this chain of benefits. As a bridge designer he is aware there are issues of long-span foot and cycle bridge design that need to be discussed.

Figure 3.1 Rope bridge near Monmouth.

Much of the demand is for cycle bridges but the topic of this paper is footbridges because both the dynamic inputs and acceptance levels of pedestrians are more onerous than those of cyclists, and cycle bridges are always available for use by pedestrians. The author's only knowledge of the dynamics of footbridges under cycle loading comes from his own experience on a 54 m span rope bridge (Fig. 3.1) over the River Wye near Monmouth, Wales. It is very lively under pedestrian loading, but under several crossings on a bicycle it was steady and stable.

There are fundamental requirements driven by the need for dynamic stability in wind and under pedestrian excitation and the solutions to these requirements affect the basic configuration and/or the alignment of the bridge. Designers stand at the threshold of an engineering adventure in which configurations are sought which meet the dynamic needs of longer spans. This adventure is in danger of being masked by the earlier architectural adventure which also side-lined the simple form of traditional footbridges. The engineering challenge is likely to result in exciting bridges whose form celebrates more fundamentally our mastery of our environment.

This paper updates and extends an earlier paper by the author [1].

3.2 CASE HISTORIES

The current situation is best told through a few brief case histories. The author's knowledge of some of these is limited but, taken together, they give an indication of the current state of the art.

3.2.1 London Millennium Bridge

This bridge has a main span of 144 m and an overall width at midspan of 11.7 m, although the width of the walkway is 4 m. It was opened on 10 June 2000 and crowds marched across enthusiastically. The bridge responded with a lateral swaying motion which, at times, increased inexorably until walking became almost impossible. The story is told in [2, 3]. The bridge was closed and the effect was investigated. It was found that data on the human response to the relevant motions did not exist. A number of experiments were instigated and the results analysed. The input to lateral deck motion imposed by a crowd depends on the response of an individual to cyclic lateral motion, the degree to which members of the crowd synchronize with the deck motion, and the phase shift between the motion and the input force. It was found that the combined effect was equivalent to an in-phase force which was proportional to the number of people on the deck and the lateral velocity of the deck. Damping is usually modelled as a force which is proportional to velocity so the effect of synchronous lateral excitation, SLE as it became known, can be thought of as negative damping. If the damping of a bridge structure, its inherent damping plus the added effect from damper units, is insufficient to match this negative damping then there is a continuing energy input into the deck and the lateral amplitude continues to increase until the pedestrian input drops. The bridge was retrofitted with a system of many dampers, sufficient for a walking density of 2 people/m^2 and has operated satisfactorily since.

3.2.2 Strasbourg-Kehl

The author entered this competition with Agirbas Weinstroer Architektur in April 2000 for a footbridge across the Rhine, linking two parts of a garden festival, with a 200 m span tied arch scheme (Fig. 3.2) in which the deck had an offset boom on outriggers. The plan form of the bridge continued the circular perimeter path across the river and the offset structure compensated for this arc. The design pre-dates the events on the London Millennium Bridge and the similarities with the Linz scheme below can only be ascribed to chance.

The competition was won by Marc Mimram as architect and engineer with a 130 m span cable-stayed scheme. A recent paper [4] shows the built version. The site

Figure 3.2 Arup/Agirbas Weinstroer entry for Strasbourg competition.

was moved to the middle of the park and the span was increased to 180 m but the basic concept with two structurally connected decks remains. One has a user width of 3 m and follows the curvature of the perimeter path, even though it is now nowhere near the perimeter, and the other with a user width of 2.5 m is straight in plan but drops towards its ends to meet the river banks. This concept results in a bridge with a structural width at midspan of 23.4 m. There is triangulated bracing in the bay between the two decks so the deck is very stiff laterally.

3.2.3   Nesciobrug, Amsterdam

The author led the engineering design for this bridge (Figs. 3.3, 3.4) [5] with Wilkinson Eyre Architects, opened in 2006. Some of its details are discussed in [6]. The span is 170 m and the user width is 5 m for combined use by pedestrians and cyclists. The concept of this bridge was developed with knowledge of the lessons learnt in London. The deck is bifurcated at each end in a way which increases the lateral stiffness. It is also curved in plan. A key question which we, as designers, had to discuss with the client was what walking density should the additional damping be designed for. The cost of providing damping is significant, and walkers are expected to be secondary users of the bridge which is principally for cyclists. The bridge has been designed for 0.5 people/m² over the full area of the deck. It has tuned mass dampers (TMDs) within its steel box-girder deck to damp lateral motions – 10 tonnes at midspan and 3 tonnes at each quarter-span position. Because the deck is curved in plan it would have been possible to replace the midspan dampers with viscous dampers at the ends acting on the in-line movements of the ends of the arch as it sways laterally in a full-length mode. However there are issues relating to the specification of viscous dampers which need to be effective under the very small movements needed to control SLE, while surviving the much larger motions due to wind. Because the procurement of TMDs is much more standard, this was the option chosen.

Figure 3.3 Nesciobrug elevation.

3.2.4   2006 Olympic Footbridge, Turin

For this footbridge (Fig. 3.5) [7, 8] with a clear span of 150 m and a walkway width of 4 m Arup was retained solely to conceive and specify a damping system for SLE. A tall steel arch is built across the line of the bridge, and the stays run from positions around the arch to positions along the deck. The deck is curved in plan and, on this occasion, the opportunity was taken to use in-line viscous dampers at the ends of the span. There were delays to the project and the bridge was only open for limited use

Figure 3.4 Nesciobrug underview.

Figure 3.5 Olympic Footbridge, Turin (Photo: Michel Denancé).

during the Winter Olympics, without the dampers. They have since been installed and the bridge is open for all users.

3.2.5 Linz

This was an open competition run by the city authorities of Linz for a footbridge/cyclebridge to cross the River Danube. A main span of 270 m was required, with a user width of 6 m. The author entered the competition with a development of the tied

Figure 3.6 Arup proposal for Linz.

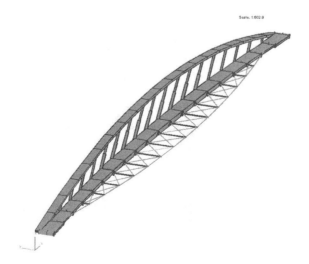

Figure 3.7 Arup proposal for Linz, showing plan bracing.

arch scheme he had entered for the Strasbourg-Kehl competition (Figs. 3.6 & 3.7). The deck structure is a 6 m wide steel box girder which increases in depth towards midspan where it is 3 m deep and is curved slightly in plan. The arch is straight in plan and is stabilised to some degree by stiff hangers interacting with the torsional stiffness of the deck. The deck follows what the author calls a "side-saddle" alignment – it has just enough curvature so that the centre of gravity of the whole deck lies under the centreline of the arch. A counterarc chord is supported on outriggers at the same level as the deck. It is sized so that the frequencies of the second and higher lateral modes exceed the SLE limit of 1.3 Hz, and the resultant centroid of the combined chord and deck lies close to the plane of the arch. To achieve this the outrigger bays need to be cross-braced. The overall width of the structure at midspan is 20 m. Even with this width it was estimated that 48 tonnes of TMD are required near midspan to control SLE in the fundamental lateral mode. This scheme was not one of 5 selected for the second stage of the competition. When a winner was declared it was also announced that there were insufficient funds available and the project was abandoned.

For the first stage of the competition there were no technical rules specified. From the images of the short-listed bridges shown on the web site most of the bridges relied only on the basic deck width to provide lateral stiffness, and they would not have met the requirements for SLE.

Although this project is not proceeding further it is interesting that a local authority is considering building a footbridge with a 270 m span. How many footbridges with such a span will be built in the next few years?

3.2.6 Weil-am-Rhein

This competition winning foot and cyclebridge (Fig. 3.8) [9, 10] connects Germany to France across the Rhine near Basel. Also called the Tri-countries Footbridge, it was

Figure 3.8 Weil-am-Rhein Footbridge.

opened in March 2007. It is a shallow tied arch with a clear span of 230 m. There is an arch each side of the deck, one in a vertical plane, and one inclined inwards, to respect an existing view. The walkway is 5 m wide at midspan and increases to 5.5 m at the ends. The springings of the arch are below deck level, quite close to the water level. The horizontal forces at the springings are resolved with diagonal members about 12 m long which connect them to the ends of the deck. The east end of the bridge is free longitudinally, but vertical and horizontal restraints at the end of the deck result in an arch which has rotational fixity in both elevation and plan at both ends.

Reference [10] assesses the pedestrian density for SLE as 0.24 people/m², but this is based on [11] and the critical density would be about a half of this value if based on [2]. A 10 tonne TMD was mentioned, but has not been provided. The author has discussed this design decision with Holgar Svensson of Leonhardt, Andra und Partner, the engineering designers of the bridge. It was anticipated that capacity pedestrian flows would be unusual in this location and, as with Nesciobrug, it has a dual role carrying both cyclists and pedestrians. Tests were conducted on the completed bridge with about 1000 people crossing and recrossing the bridge. Under natural walking conditions no SLE was observed. With marshals to speed up the walking speed SLE was induced during the tests. It was concluded that these tests corroborated the decision not to provide additional damping.

3.3   DESIGN REQUIREMENTS FOR A LONG-SPAN FOOTBRIDGE

The designer needs to address four dynamic effects:

- Divergent aerodynamic instability
- SLE
- Vortex excitation
- Anxiety/comfort under vertical motions

Of these only the first has hard, engineering criteria to be satisfied. All the others are, to some extent, soft issues which need to be understood within the human context [12].

3.3.1   Divergent aerodynamic instability

Aerodynamic stability under the maximum wind speed is probably the main challenge when designing a long-span footbridge. This is a big subject that is not addressed here. It has to be studied in a wind tunnel although, as mentioned in 3.4 below, virtual wind tunnels are being developed.

3.3.2   Synchronous Lateral Excitation

This topic has been mentioned several times and it needs to be fully understood. A simple design study is presented below. Since the studies made for the retrofit of the London Millennium Bridge there has existed a procedure for checking for SLE [2, 3].

It evaluates the critical number of people who can initiate SLE. On some occasions during the trials the bridge was stable with greater than the critical number of people. It was as if the system was waiting for a trigger to initiate the instability. The procedure does not address the issue of what walking density to design for. The author provisionally suggests 1.5 people/m² for a footbridge. There are a number of teams developing guidelines that cover this [13, 14]. It was suggested that the design walking density should depend on how often a particular bridge might receive crowd loading.

Of the five built case histories reported above three have additional damping to meet the needs of SLE, Strasbourg is unusual in having a deck width which is 13% of its span and Weil-am-Rhein has plan fixity at its ends, and a deck which narrows towards mid-span. Weil-am-Rhein is the interesting one because of the conscious decision not to provide additional damping. The soft decision that has to be made, in conjunction with the client, is what walking density to design for. And the trigger issue – the fact that the bridge can remain stable under a crowd sufficient to cause instability – may significantly reduce the number of times that an undamped bridge, such as Weil-am-Rhein, will become unstable. There is some evidence that a lateral amplitude of about 10 mm or 15 mm is sufficient to trigger synchronisation and the consequent instability. This is a contentious issue because the role of synchronisation in SLE is disputed [15].

3.3.2.1 SLE design study

However the vertical loads are carried it is often the case that the transverse loads on a footbridge are carried back to the supports with simply supported beam action in the deck, and this is the basis of this study. For a range of spans the total mass of TMDs is found which is sufficient to stabilise a walking density of 1.5 people/m². The walkway is assumed to be 4 m wide which is sufficient for significant pedestrian flows, or light flows of cycles with pedestrians. Fig. 3.9 shows the deck section assumed for the study. Its breadth is just enough to carry the walkway and its parapets. The form of the deck, a steel box-girder with 40 mm side plates, has been chosen recognising that lateral stiffness is an important requirement. It spans simply-supported between the

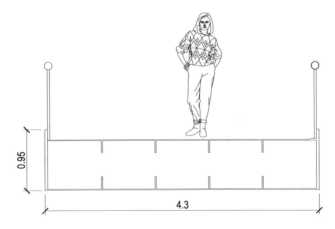

Figure 3.9 Assumed deck section.

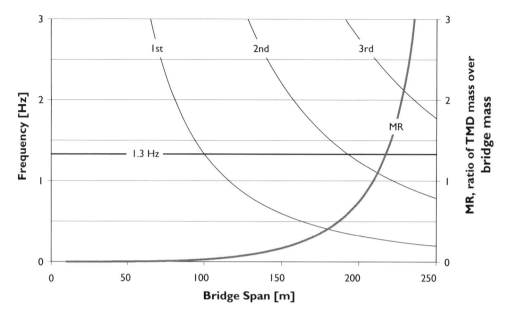

Figure 3.10 Mass Ratio varying with span.

ends of the span and it is assumed that the total mass of the deck is 15% more than that of the basic steel section.

Fig. 3.10 shows the effect of varying the span. The frequencies of the first three lateral modes are plotted, together with the limiting frequency of 1.3 Hz from [2]. Also plotted is the mass of TMD needed for the first mode, expressed as a ratio of the mass of the bridge. The required damping is calculated in accordance with Equation 9 of [2], assuming the inherent damping in the structure is 0.3% of critical. The sizing of the TMDs follows guidance in [16]. The TMD is assumed to operate with an efficiency of 0.75 and the optimal mass is calculated using Equation 6.

The graph shows that mass requirements of TMDs increase very rapidly for longer spans, and so they are only useful over a limited span range. In practice it is difficult to achieve an additional damping ratio of more than about 15% using TMDs. This is equivalent to a mass ratio MR of about 0.10, which In this case limits the span to about 130 m. Longer spans need added lateral stiffness.

3.3.3   Anxiety and comfort under vertical motions

These are two issues here, anxiety and comfort, which have the same effect on the bridge design – defining limits on the motion – but the human issues are very different.

Nervous people may find themselves unable to use a bridge which has a particularly lively response to foot fall, or is subject to noticeable vertical motions caused by vortex excitation in a moderate wind. This is a psychological response to the perception of a bridge as being part of *terra firma*. The same motions would probably not alarm the same people if experienced in a car, a boat or a plane. As longer span footbridges become more common users will become used to the fact that a footbridge is something which moves. This can be assisted by information boards and press releases. However there clearly are limits to acceptability and it is difficult to make judgements on this matter. Data is needed on the experiences of users.

The principal factor which affects the comfort of users of footbridges is probably the weather. When considering any provision against rain it is usual to follow the precedence of the public realm on either side of the bridge. If the approaches are uncovered then the footbridge should also be uncovered. However, in the case of wind there is a different issue. Footbridges are usually more exposed to wind than their approaches and there may be a case for providing raised solid or part-solid parapets to provide some shielding. Unfortunately such parapets can make it much more difficult to solve the aerodynamic problems. Design studies are needed to identify the extra costs to provide the additional stability required for wind shielding.

Deck motions generated by pedestrians or by wind are increasingly being included in comfort criteria for the design of bridges. The author is sceptical about this as a comfort issue, and suspects that it is being confused with the anxiety issue. The author is also concerned that criteria for these soft issues will be specified which have a disproportionate effect on the cost of the bridge, and hence could affect the decision to build the bridge. Before any criteria are fixed, design studies should be carried out to find how the criteria are likely to affect the resulting designs.

Usually the "comfort" limits are expressed as vertical accelerations experienced by those standing on the bridge although the visible amplitude of motion seen by those about to step onto the bridge is also a factor which causes anxiety. It should be remembered that judgements on human criteria need to made in the local context, and they may vary from bridge to bridge.

Even if there were agreement on acceptable accelerations, there is no agreed procedure for assessing the vertical accelerations of a footbridge to pedestrian excitation. The problem is modelling its response to groups of people. There is some evidence that people provide significant additional damping, so the maximum response occurs under a relatively small group of people – possibly 6 or 8. Hence the problem cannot be studied with the usual separation of structure and load, because the load provides some of the characteristics of the structure.

3.3.4   Vortex excitation

This is a resonance phenomenon that occurs at a specific wind speed for each bridge. It usually produces a vertical cyclic motion of the deck. Nesciobrug was tested in a wind tunnel with a 1:50 aeroelastic scale model. Vortex induced motions at midspan of about ± 60 mm were predicted at 10 m/s and these have been observed in the completed bridge. The depth of the deck varies from 0.69 m at the ends of the main span to 1.95 m at midspan. This variation reduces the response because the vortex frequency varies with the depth of the deck and the wind speed. At any wind speed only a portion of the span is being excited at its modal frequency.

Because there was an issue about the parapet mesh for Nesciobrug getting partially blocked by snow the wind tunnel tests were run twice, with open parapets and with 50% blockage. It is interesting that the case with 50% blockage gave

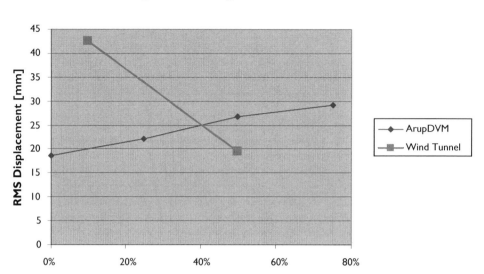

Figure 3.11 Vortex amplitudes.

a lesser response than the open case. This suggested that there might be a form of deck section which provides some degree of wind-shielding without significant aerodynamic penalty. More recently within Arup we have written Discrete Vortex Modelling (DVM) software which acts as a digital wind tunnel. The Nesciobrug sections have been investigated and the results are shown in Fig. 3.11. DVM indicates an opposite result with the vortex response increasing with degree of blockage. When a digital model disagrees with a wind tunnel how do we know which is right?

There are two factors to consider when deciding what motion is acceptable. The first is fatigue. Because extreme winds are so much stronger than frequent winds, fatigue is not usually a concern under wind loads. However this is a fluctuating action under a steady wind. Climate data is needed to find the time that the wind speed will be within a band of speeds centred on the critical speed. This criterion is unlikely to govern the design of the deck.

The second is the anxiety/comfort issue discussed above. Vortex excitation is disconcerting because it appears to be disproportionate to the wind speed. This is where specific information can be very reassuring. The public should be told the maximum movement expected, the wind speed that it is likely to occur at, and the fact that the motion reduces at higher speeds. This could be included on a notice board with general information about the bridge. It need not be in a large typeface. If the information is there it will be noted and in time will become generally known.

3.4   DESIGN STRATEGIES

Dynamic excitations can be controlled with:

- additional stiffness
- additional damping
- additional mass.

The designer has to have a good understanding of dynamic behaviour to see what action is most likely to lead to a satisfactory and economic solution. The strategy needs to address the four dynamic effects together.

Mass can be very useful when controlling higher frequencies and short spans, but it is not appropriate for longer spans.

Stiffness is needed to provide aerodynamic stability under extreme wind conditions, although it should be stated that aerodynamic damping also contributes to stability. There may be structures where viscous dampers could be employed to control aerodynamic instability under extreme winds, but the author does not know of dampers being used for this.

Damping with TMD can be used to control vortex excitations, although variations of the deck section should be studied first in the wind tunnel to solve the problem at source, as far as possible.

As has been explained above, SLE acts as negative damping and so it is natural to control it with positive damping. The case histories also show that stiffness is an important part of the control strategy. At Linz only one mode required damping because the other modes had been taken above the limit frequency by using additional structure to provide stiffness. Weil-am-Rhine is a reminder that it may be acceptable to specify quite a low density of walkers, but it should be noted that the strategy here also included additional stiffness in the form of plan fixity at the ends and, probably, a deck which is wider than is needed for the usage. Widening the deck is an inefficient way to increase the walker density because the excitation is also being increased, but at a slower rate than the resistance. There is a definite benefit in using outrigger structure which increases stiffness without increasing the number of people, and the author believes that the whole footbridge configuration for Linz has much to commend it. If choosing a reduced walker density there should be an awareness of the cost of the additional structure needed to control the full density of 1.5 people/m². In consultation with the client a designer might design a structure with sufficient stiffness for a compliant design when used with a practical size of TMD, and then omit the TMD from the initial construction. If it is needed it can be added later.

3.4.1   Dampers

The detailed specification of damping units is a specialist topic which is outside the scope of this paper. However it is necessary for the bridge designer to understand how different dampers fit into design strategies, and to recognise the conditions in which a simple, cheap and reliable damper can operate effectively. Some case histories and guidance are reported in [16, 17]. To control SLE dampers have to act on very small movements, and special dampers are needed with a high specification.

Dampers act on the modal behaviour of the structure. If the damping level is quite low the mode shapes are not much affected. If a high level of damping is applied, the forces in the dampers will disrupt the behaviour and alternative modes will develop.

All dampers dissipate energy which is input to them through cyclic motion. Most such dampers act through the viscosity of a fluid being forced through a constriction. They are called viscous dampers and are effective at any frequency. Dampers can be used in a footbridge in three ways:

3.4.1.1   Internal Reactive Dampers

These are placed between parts of the structure which move relative to each other under the modal action which needs to be damped. In many forms of bridge construction, such as steel or concrete box girders, there are no significant movements between parts. Where there are relative movements they are usually very small and ancillary structure is needed to transfer the two sides of the movement to a location where the damper can act across a gap. This was one of the solutions used for the London Millennium Bridge. Dampers for such situations need to be carefully specified because they typically operate on movements of less than a millimetre.

In cable supported bridges there may be opportunities to exploit relative movements between the cables and the deck, or between different parts of the cable system.

3.4.1.2   External Reactive Dampers

There are likely to be large relative movements between the modal displacements of the deck and the surrounding landscape. The problem here is one of distance. It would be possible to attach a taut thin wire from the deck to an anchor structure on land. One end of the wire would be tensioned with a spring which has sufficient travel to follow the maximum displacement of the deck. The damper would be fixed across the ends of the spring. It would need to operate effectively under the small movements and forces which characterise bridge decks with stable crowds on them, and yet it must also be able to survive the much larger motions and forces imposed by high winds. Damper systems which release the forces under large movements are being developed.

It is a matter of semantics whether a damper which acts on the relative movements between the deck and the substructure is internal or external. The author feels they are best described as external. Again there are examples on London Millennium Bridge. There are long telescopic cylinders placed diagonally between the top of the piers and the deck.

As already mentioned the full-length mode of a deck curved in plan can be damped with an in-line viscous damper at one end, although this does not work for antisymmetric modes.

3.4.1.3   Tuned Mass Dampers (TMDs)

Where there is no relative movement, a single movement can be used in conjunction with the inertia of a mass. If the supports of the mass are tuned to the frequency of the mode to be damped, this can be effective with quite a small mass. However, if there is more than one problem mode, separate TMDs are needed for each mode. Low frequency modes require larger masses. TMDs have been used successfully on a wide range of structures. On a longer span footbridge with three or more modes to damp, it would require many dampers, and a significant additional mass on the bridge.

3.4.2 Shape

This has been mentioned above but, for completeness, it is repeated here because it is an important strategy. The deck section should be studied in a wind tunnel to minimise vortex excitation, as well as to achieve stability under maximum wind speeds. Also, as spans become longer, it will be difficult to limit vortex motions to an acceptable level without varying the depth of the deck section.

3.5 CONCLUSIONS

The publicity attracted by the London Millennium Bridge has drawn attention to the need to add TMDs on longer span footbridges to control SLE. From studies it is now clear that, in general, additional transverse stiffness is often also needed, and this affects the basic configuration of the bridge.

Many of the design criteria for dynamic actions relate to human issues and as much feedback as possible is needed on what is found to be acceptable or unacceptable in practice. Designers must recognise that the human context may differ between projects and the criteria should be reassessed each time. There is a concern that the economic viability could be affected by unduly onerous specifications for human issues.

The dynamic issues need to be addressed at the very beginning of the project. Some of the design strategies require curved plan alignments and so the location of the bridge might be affected.

REFERENCES

[1] Low, A., "The design of long-span footbridges". *First International Conference on Advances in Bridge Engineering*. Brunel University, London. 2006.

[2] Dallard, P., Fitzpatrick, A.J., Flint, A., Le Bourva, S., Low, A., Ridsdill-Smith, R.M., and Willford, M., "The London Millennium Footbridge", *The Structural Engineer*, Vol. 79, No. 22, 20 Nov 2001, pp. 17–33.

[3] Dallard, P., Fitzpatrick, A.J., Flint, A., Low, A., Ridsdill-smith R.M., Willford M., and Roche M., "London Millennium Bridge: Pedestrian-induced lateral vibration", *ASCE Journal of Bridge Engineering,* 2001, **6**, No. 6, pp. 412–417.

[4] Morgenthal, G., Kovacs, I., and Saul, R., "Analysis of Aeroelastic Bridge Deck Response to Natural Wind." *Structural Engineering International*. IABSE, 4/2005.

[5] Low, A., and Ichimaru, Y., "Nesciobrug, Amsterdam." *The Structural Engineer*, IStructE London. 6 March 2007.

[6] Low, A., and Ichimaru, Y., "Nesciobrug, Amsterdam." IABSE Symposium, Weimar 2007.

[7] Dutton, H., Poronne, I., Bucci, P., Tarditi, S., and Soldani P., "2006 Olympic Footbridge, Turin, Italy." *Footbridge 2005*, Venice 2005.

[8] Dutton, H., "Passerella Olimpica à Turin", bulletin Ouvrages Metalliques No. 5 2008, OTUA, Paris.

[9] Strobl, W., and Haberle, U., "A pedestrian arch bridge with a span of 230 m". IABSE Symposium, Weimar 2007.

[10] Strobl, W., Kovacs, I., Andra, H.-P., and Haberle, U., "Eine Fußgängerbrücke mit einer Spannweite von 230 m". Stahlbau 76 (2007) pp. 869–879, Ernst & Sohn, Berlin.

[11] Fujino, J., et al. "Synchronisation of human walking during lateral vibration of a congested bridge". US-Japan Workshop, Reno, Nevada, 1990.

[12] Low, A., "Soft issues in the design of long span footbridges and cycle bridges". Footbridge 2008, Porto.

[13] Schlaich, M., "Guidelines for the Design of Footbridges." *Footbridge 2005*, Venice 2005.

[14] Barker, C., Mackenzie, D., Mcfadyen, N., Deneuman, S., Ko, R., and Allison, B., "Footbridge Pedestrian Vibration Limits, Parts 1–3." *Footbridge 2005*, Venice 2005.

[15] Macdonald, J.H.G., "Pedestrian-induced vibrations of the Clifton Suspension Bridge, UK". ICE Proceedings. Bridge Engineering 161 Issue BE2. ICE, London, 2008.

[16] De Donno, A., Powell, D., and Low, A., "Design of damping systems for footbridges – Conceptual framework" *Footbridge 2005*, Venice 2005.

[17] Powell, D., De Donno, A., and Low, A., "Design of damping systems for footbridges – experience from Gatwick and IJburg." *Footbridge 2005*, Venice 2005.

Chapter 4

Footbridges, numerical approach

Krzysztof Zoltowski
Gdansk University of Technology, Gdansk, Poland

SUMMARY

The paper is a review of numerical simulations of pedestrian bridges. The subject is divided into three issues.

1   FEM structural models of pedestrian bridges
2   Live loads on pedestrian bridges
3   Response of the footbridge – numeric simulations,

   – deck vertical vibrations under vertical excitation,
   – cable stayed footbridge – vibrations of stays.

Selected theoretical bases and formulas are presented in this paper in a condensed way to explain principals and problems which were important for the author in practice.

Keywords: footbridge, dynamics, vertical pedestrian load, response, guidelines for designing.

4.1   INTRODUCTION

All we know that a modern footbridge needs modern techniques of analysis. Thanks to footbridge conferences we are participants of a big discussion about theoretical evaluation, comfort criteria and control of vibration. A very important issue of this discussion is a numerical simulation of response of a footbridge under dynamic excitation. The main and the most spectacular problem is a dynamic action on the bridge. Periodic load from pedestrians and wind can accelerate a bridge to the level which can be dangerous for the structure itself or at least too large to be tolerated. As we are not interested in a replay of London Millennium Bridge story, a numerical efficient procedure for static and dynamic analysis of structures is the key element on the way to succeed in constructing a modern footbridge.

4.2   FEM NUMERICAL MODEL

The general dynamic problem is given by the well known discreetised differential equation:

$$y''M + Cy' + Ky = P(t) \tag{1}$$

where:

y	– displacement vector
M	– mass matrix
C	– damping matrix
K	– stiffness matrix
$P(t)$	– external loading

K and M represent the structural part of our numerical model. Point masses, three – dimensional prismatic bending beams, truss and cable elements, spring elements, boundary and flex elements, shell elements, 3D – solid elements, all these Final Elements are available in most commercial FEM systems. However, basic knowledge is necessary to maintain with FEM program to avoid mistakes. A good mechanical model of an engineer structure with efficient final elements, proper discreetisation, boundary conditions, realistic material model and loads is the key aspect in designing.

4.2.1   Structural model – K and M from formula (1)

In most cases a well-known beam FEM model can solve all practical static and dynamic problems. However we should check if a FEM beam element used for computation involves shear deformations and active width. Shear deformations can have an influence in total dynamic deformation of the structure. Active width as a reduction of flange or deck coupled with a beam is necessary in dimensioning of cross sections. We have to remember that active width parameter can reduce the mass of an element, too. Therefore before a computation of a final problem a careful study of software manual is obliged and parallel check of the new procedure on a simple example with a known theoretic solution is strongly recommended. Fig. 4.1 presents a simple continuous beam system with two steel girders and an orthotropic deck.

This simple beam model (Fig. 4.1) satisfies us in 100%. Calibration with real structure was not necessary. A pure FEM beam model without active width factor gives 95% accuracy in deflections and eigen frequencies (proved by site test). It means

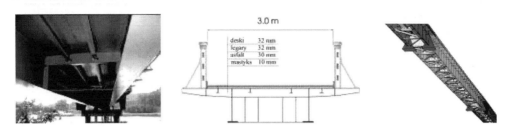

Figure 4.1 Pedestrian steel bridge. Cross section and FEM SOFiSTiK beam model.

Figure 4.2 Millennium Bridge in London. Designed by Ove Arup.

Figure 4.3 Footbridge in Minden. Designed by SBP.

that K and M from formula 1 is defined properly. A separate and key problem comes with definition of C but to this matter we will come back later.

Beam-truss-cable FEM model can properly simulate modern and spectacular footbridges. Fig. 4.2 presents a beam model of Millennium Bridge in London and Fig. 4.3 shows a Minden footbridge. In both bridges the main structural idea comes from a ribbon or suspension system.

A suspension cable in the structural system creates theoretical and numerical problems in FEM structural model. That is why, before we start dynamic analysis we have to solve non-linear geometrical static problem to find the final geometry and internal stress in structure under the dead load. If we represent in our FEM model suspension cable by truss or cable elements we can expect numerical problems due to the small initial stiffness of the structure. Therefore, an advance study of this problem is necessary to pass initial instability. After that we can start the simulation of dynamic effects.

New, powerful FEM software gives us the chance to simulate easy (with a small effort) our structure more exactly using continuous flat shell elements and 3D volume elements. The combination of beam, shell and volume elements was used to analyze the footbridge over Woloska street in Warsaw (Fig. 4.4)

Detailed meshing (Fig. 4.5) can give the answer to many important questions like collaboration of an orthotropic deck with the main girders or effect of not typical anchor of stay in the main girder (Fig. 4.5). Unfortunately for dynamic simulations this type of modelling is very time consuming and not necessary regarding the expected results. However, an advanced FEM structural model can be helpful in verification of a simplified solution. On the base of modal analysis the simplified system can be calibrated to the advanced one.

Fig. 4.6 presents a simplified beam model of a footbridge. The structural model of a span is reduced to one beam element in cross section but stays are divided into several elements. This model can simulate excitation of a span, pylon and stays.

Figure 4.4 Footbridge over Woloska street in Warsaw. Designed by Transprojekt Gdansk. General view, back stay anchor block, stay anchor in main span.

Figure 4.5 FEM SOFiSTiK model of Woloska footbridge.

Figure 4.6 Simplified beam model of Woloska footbridge with advanced multi element stays.

The important results of analysis can be transferred to the advanced FEM model and computed as an equivalent static problem.

On the base of a numerical model explained above, a designer can evaluate the superstructure under static and dynamic load. Mostly, dynamic analysis can be done on the base of a modal superposition and time step. An advantage of this method is the efficient reduction of time in computation based on reduction of degrees of freedom in a numerical problem. Unfortunately this type of analysis is possible for pure linear problem or for small vibrations around the stay of equilibrium. In case of cable or suspension structures when precise cable action is important the full non-linear time-step direct analysis developed by Newmark and Wilson is necessary.

4.2.2 Damping – C from formula (1)

A very important and most unknown element of equation (1) is a damping matrix C. It has to be considered in dynamic analysis because it reduces dynamic effects on the footbridge. On the base of many experiments we know damping properties of basic materials used for bridge constructing. Unfortunately ready-made structures are combined from several materials and damping parameters of readymade structure are mostly bigger than these developed for pure materials. This phenomenon comes from the nature of damping which we can divide into three aspects:

– internal damping as an effect of dissipation of energy in pure material,
– external damping as a result of not perfect structural connections, failures or boundary conditions.
– specially created structural elements (dampers) prepared for dissipation of energy.

Unfortunately practically used in analysis, damping characteristic is based on experience with similar structures. Only specially created structural dampers can be defined more precisely.

In engineering practice we have two most popular damping parameters based on viscotic damping Voigt hypothesis:

For SDOF model

$$\text{Damping ratio} \quad \xi = \frac{c}{c_{critr}}, \quad c_{critr} = 2m\omega, \quad \xi = \frac{c}{2m\omega} \tag{2}$$

$$\text{Logarithmic damping} \quad \Lambda = ln\frac{y_j}{y_{j+1}}, \, y_j - \text{amplitude of vibration} \tag{3}$$

$$\Lambda = \frac{2\pi\xi}{\sqrt{1-\xi^2}} \tag{4}$$

For MDOF model damping formula developed by Rayleigh is implemented:

$$y''M + Cy' + Ky = P(t) \quad C = a_0 M + a_1 K \tag{5}$$

Damping matrix is a proportional combination of stiffness matrix and mass matrix.

$$\text{If } \xi_n = \frac{C_n}{2M_n\omega_n} \quad \text{and} \quad \xi_n(M) = \frac{a_0}{2\omega_n}, \; \xi_n(K) = \frac{a_1 K}{2M_n\omega_n} = \frac{a_1\omega_n}{2} \quad \text{than}$$

$$\xi_n = \frac{a_0}{2\omega_n} + \frac{a_i\omega_n}{2} \tag{6}$$

Formula (6) gives a damping ratio of n-eigen form of MDOF structure system. Using a commercial FEM program the designer has a possibility to define factors a_0 and a_1 to implement damping to numerical model. This factors are used by FEM code to simulate damping by modification of K and M matrix. However we have to remember that damping ratio defined in this way is proper to exactly one eigen form. Therefore before dynamic analysis we have to identify the main eigen form corresponding to excitation. Finally it is easy to check if our damping parameters are sufficient. This can be done by analyzing the response of our structural model to impulsive load. Real damping in our model can be evaluated from history of vibration of representative point of the structural model.

If we do not have specially dedicated dampers in our superstructure, we can estimate damping ratio by studying similar readymade structures. That's why it is very important to create a data bank of information concerning dynamic tests on site and evaluated damping factors.

4.2.3   Loads – P(t) from formula (1)

All static loads concerning dead and live weight for footbridges are defined precisely in every national code. Therefore only dynamic action of pedestrians on the footbridge will be presented in the following part of the paper.

4.2.3.1   Single pedestrian

On the base of a laboratory test made on the fitness treadmill with electric steering placed on force cells (fig. 4.7) a load function of a single pedestrian was developed [1] (fig. 4.8 and formula (7))

$$F_b(t) = BW\left\{1 + A\left[\begin{array}{l} sin(\omega_b t + \varphi) + 0,25sin(2\omega_b t + \pi + \varphi) \\ +0,25sin(3\omega_b t + \pi + \varphi) \end{array}\right]\right\} \tag{7}$$

$$A[kN] = 0,4\frac{\omega_b}{2\pi} + 0,6BW - 0,84 \tag{8}$$

BW body weight [kN]
ω_b pacing rate
φ phase shift

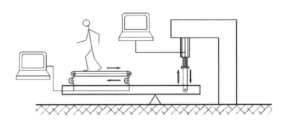

Figure 4.7 Fitness treadmill on the vibrating lever.

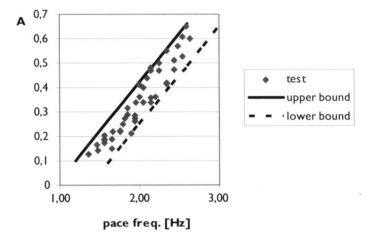

Figure 4.8 First harmonic amplitude of load function (Eq. 7).

4.2.3.2 Interaction of a pedestrian and a vibrating deck (locking effect)

Locking effect during a walk on the vibrating deck has an origin in natural adapting mechanism which reduces forces in human body [1]. This effect can be reached if the centre of a human mass (stomach) gets as small acceleration as possible. The lock in effect is a synchronization of the frequency of steps and the frequency of the deck coupled with the phase shift. The analysis of the test results and movies shows that the center of human mass is moving up when the deck is moving down. That means that locking effect reduces dynamic part of human induced load where the simple condition for phase shift is fulfilled $\varphi = \pi$.

The simplest model of the system can simulate the phenomenon of lock in effect (fig. 4.9). M_h represents the mass of a pedestrian, $h(t)$ defines vertical movement of human mass while walking, M_b the mass of the bridge.

$$y_b'' M_b + C y_b' + K y_b + (b'' + y_b'') M_h = 0 \qquad (9)$$

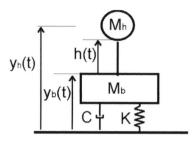

Figure 4.9 A simple model with a single degree of freedom.

$$y_b'' M_b + Cy_b' + Ky_b + h'' M_h + y_b'' M_h = 0 \tag{10}$$

$$h'' M_h = F_h \qquad\qquad F_h(t) \text{ is defined in (7)}$$

$$y_b'' M_b + Cy_b' + Ky_b + F_h + y_b'' M_h = 0 \tag{11}$$

$$y_b'' M_b + Cy_b' + Ky_b + L_h = 0 \qquad L_h = F_h + y_b'' M_h \qquad M_h = \frac{BW}{g} \tag{12}$$

Equation (12) specifies a SDOF system with nonlinear external load $L_h(t)$. If we ignore the static part of a human load, the load and the bridge frequency are equal and the load function is shifted in the phase by π (lock in effect) and

$$y_b(t) = -U(t)sin(\omega_b t) \quad \text{than:}$$

$$L_h(t) = sin(\omega_b t)\left[BW \times A - U(t)\omega_b^2 M_h \right] + BW \times A \begin{bmatrix} 0,25sin(2\omega_b t) \\ +0,25sin(3\omega_b t) \end{bmatrix} \tag{13}$$

if $U(t)\omega_b^2 M_h - BW \times A = 0$ there is no load anymore (the second and the third harmonic is not so important) and the final amplitude of deck deflection can be estimated as:

$$U_g = \frac{BW \times A}{\omega_b^2 M_h} \qquad U_g = \frac{M_h \times g \times A}{\omega_b^2 M_h} \qquad U_g = \frac{g \times A}{4\pi^2 f^2} \tag{14}$$

This situation shows that thanks to a lock in effect pedestrians can accelerate the structure to the finite value. If we ignore damping in (12) and 2-nd and 3-rd harmonic in (7) the final acceleration of the deck with the first natural frequency of 2 Hz (an average pedestrian mass of 75 kg, A = 0,41) is ~4,0 m/s² and amplitude of deck deflection is ~0,0255 m.

Formula (14) was verified with success by a laboratory test presented in [1].

Finally, we can assume that a pedestrian load on the lively footbridge can be defined as:

$$L_h(t) = BW \left\{ 1 + A \begin{bmatrix} sin(\omega_b t + \varphi) + 0,25sin(2\omega_b t + \pi + \varphi) \\ + 0,25sin(3w_b t + \pi + \varphi) \end{bmatrix} \right\} + y_n'' M_h \tag{15}$$

Were y_n'' is an acceleration of structural point of our deck just under the moving load and φ is an actual phase shift.

This load function can be implemented to FEM model. For that we have to trace deflections of every structural point during procedure and redefine load function during every time step. Depending on numerical model of footbridge a group of single loads (15) or a global load function (function (15) multiplied by number of pedestrians) can be implemented as a moving load on the bridge. In this way one can simulate a group of synchronized pedestrians.

Using a random generator of a phase shift, frequency and BW parameter. Several load functions can be considered together for a more sophisticated simulation. Such procedures can be done using SOFiSTiK FEM code.

4.2.3.3 Pedestrian flow

Regarding a general problem of pedestrian flow on the bridge and a density of crowd a simple laboratory test was proceeded. The participants of the test were asked to walk naturally but as closely as possible. That is why the result given in fig. 4.10 should be interpreted as an upper bound of a real situation. On the base of the trend function given in fig. 4.10 a designer can estimate a number of pedestrians on the deck and check carrying capacity limit state for synchronised flow of pedestrians on the bridge using formula (16). Equivalent uniformly distributed load $L(t)$ is calculated as a multiplication of standard load function (7) by number of people on the bridge.

$$L_s(t) = \frac{N \times BW\{1 + A[\sin(\omega_b t) + 0{,}25\sin(2\omega_b t + \pi) + 0{,}25\sin(3\omega_b t + \pi)]\}}{F_b} \qquad (16)$$

N – number of people on the bridge (can be calculated according to fig. 4.10)
F_b – total deck area

In case of not synchronized pedestrian flow a simple formula (17) developed by Matsumoto [3] can be used for evaluation of number of synchronized pedestrians on the bridge.

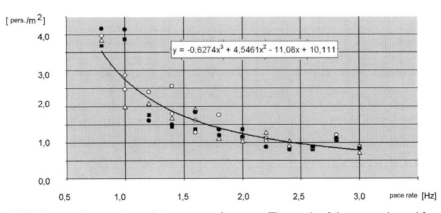

Figure 4.10 Density of a crowd in relation to step frequency. The result of the test and trend function.

$$M = \sqrt{N} \tag{17}$$

M – number of synchronized pedestrians in group of N people.

A designer can check comfort criteria for natural flow of pedestrians on the bridge using formula (18).

$$L_{ns}(t) = \frac{BW\{N + M \times A[\sin(\omega_b t) + 0,25\sin(2\omega_b t + \pi) + 0,25\sin(3\omega_b t + \pi)]\}}{F_b} \tag{18}$$

4.2.3.4   *Purposeful action on the bridge*

No one can imagine that there is a lively footbridge which was not excited by purposeful action of people. In this case danger comes from several reasons:

- man who wants to excite bridge naturally act according to natural frequency of the span,
- sensation of vibration gives motivation to further action,
- mostly excitation is proceeded by a group of people and collective action increase a filling of safety.

This is a reason why modern lively footbridge has to be checked in case of such action treated as a check of carrying capacity.

Jumping

Jumping is a human activity which create big ground reactions. But for pedestrian bridge more important is a cycling load than single incidental high impact.

Several tests with jumping people on the vibrating platform (fig. 4.7) were proceeded and we can assume that this kind of activity can not be treated as a danger. A reason is in the type of motion. As it is explained in [4] and presented on fig. 4.11 a very important phase of jumping is a fly. During this phase man has no contact with the deck and he loose synchronization with bridge motion. On fig. 4.12 representative

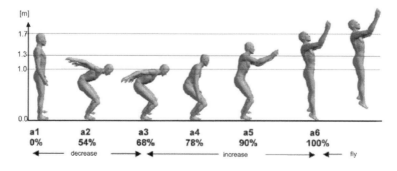

Figure 4.11 Motion in phases during jumping [4].

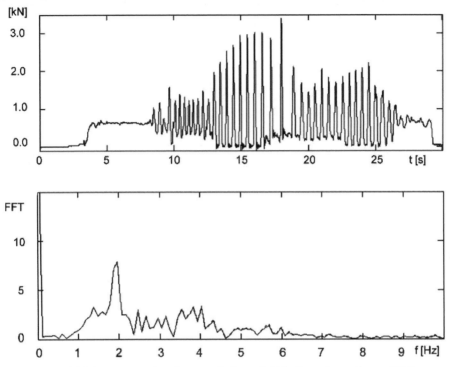

Figure 4.12 Jumping on the vibrating platform (*2 Hz*). Ground reaction and FFT.

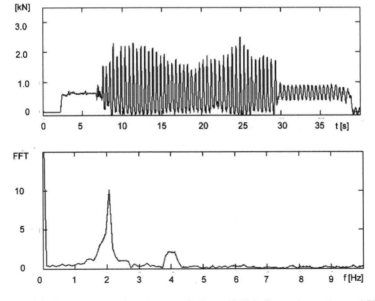

Figure 4.13 Crouching on the vibrating platform (*2 Hz*). Ground reaction and FFT.

results are presented. Exception of this statment can be a pop music concert when synchronized jumping can be a danger.

Crouching

Laboratory tests and observation of real situations are pointing that crouching is an activity which can effectively accelerate a footbridge. To compare the effect of crouching to jumping a test on vibrating platform was proceeded (fig. 4.13). One can easily notice that the amplitude of vertical load from crouching is smaller than in jumping but with crouching there is a full synchronization of a man and vibrating deck.

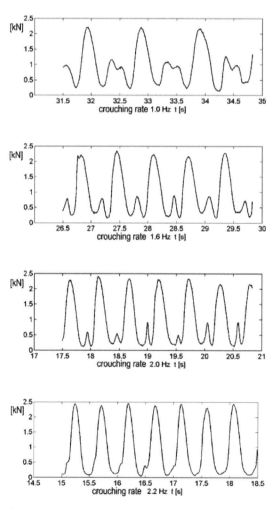

Figure 4.14 Crouching – experimental load function.

Phenomena of crouching was recognized as important dynamic activity and development of theoretic load function for crouching was a purpose of farther investigations. Fig. 4.14 presents experimental load function from crouching with different rate. Fig. 4.15 presents the same experimental function for one crouch.

$$F(t) = BW\left[A\left(cos(\omega|t| - \phi)\right)e^{-|t|\nu} + 1\right] \tag{19}$$

Comparing *RMS* of experiment data with theoretic function (19), after statistic analysis, following parameters were developed:

BW body weight, $BW = 0,75\ kN$
A dynamic factor, recognized $A = 1,3$
ω recognized $\omega = 2\pi f$, $f = 2.4\ Hz$ – constant value
φ phase shift recognized $\phi = 0.25$
ν damping $\nu = \sqrt{16T}$
T time period
t time $t\ \varepsilon <-T/2,\ T/2>$

$$F(t) = 0.75\left[1.3\left(cos\left(4.8\pi|t| - 0.25\right)\right)e^{-|t|\nu} + 1\right] \tag{20}$$

Finally theoretic load function for crouching from one man was formulated (20) and presented on fig. 4.16. Consequently a load from *M* people can be treated as a simple multiplication of formula (20).

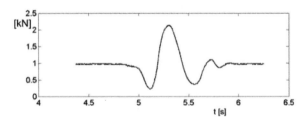

Figure 4.15 Crouching – one period.

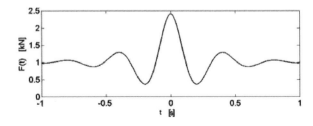

Figure 4.16 Crouching – one period. Theoretic function (13).

4.3   RESPONSE OF THE FOOTBRIDGE – NUMERIC AND SIDE TEST RESULTS.

Footbridge over Dunajec River in Sromowce Nizne, Poland

The cable stayed footbridge crossing the border between Poland and Slovakia (fig. 4.17) with glued-laminated wooden span and steel pylon [6]. Because of a relatively big main span (90 m) and narrow deck the dynamic simulation of pedestrian action was preformed in the early stage of designing using the procedure explained above in the paper. Afterwards the site test was proceeded on the ready-made bridge [7]. Basic dynamic characteristic of the bridge is presented in fig. 4.18.

Several site tests were executed on the bridge. Four of them are presented below:

- synchronized walking of 12 pedestrians with pace rate 1.31 Hz – test SW-12f_1
- synchronized walking of 12 pedestrians with pace rate 2.25 Hz – test SW-12f_2
- synchronized walking of 3 pedestrians with pace rate 2.25 Hz – test SW-3f_2
- synchronized crouching of 12 pedestrians with rate 1.31 Hz – test HC-12f_1
- synchronized crouching of 12 pedestrians with rate 2.25 Hz – test HC-12f_2

The results are presented in tab. 4.1 1 and fig. 4.19 till 23.

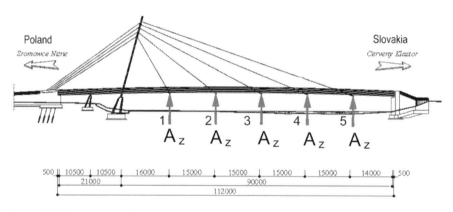

Figure 4.17 Footbridge in Sromowce Nizne, A$_z$-points of acceleration measurement devices.

Figure 4.18 First two vertical eigenforms, f_1 = 1.31 Hz, f_2 = 2.25 Hz. Developed on site damping ξ = 0.0069 (LTD = 0.0433).

Table 4.1 Sromowce footbridge. Results of site test and FEM simulation.

Test	Site test results [m/s²]	FEM simulation [m/s²]
SW-12f1	0.46	0.49
SW-12f2	1.88	1.5
SW-3f2	1.612	1.579
HC-12f1	3.55	3.30
HC-12f2	3.89	9.25

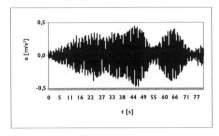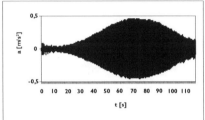

Figure 4.19 Footbridge in Sromowce, Test SW-12f$_1$. Site test (left), FEM simulation (right) node no. 4 (fig. 4.17).

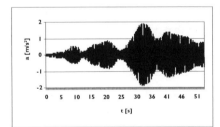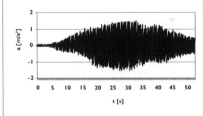

Figure 4.20 Footbridge in Sromowce, Test SW-12f$_2$. Site test (left), FEM simulation (right) node no. 4 (fig. 4.17).

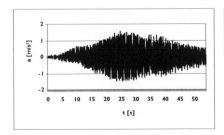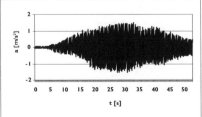

Figure 4.21 Footbridge in Sromowce, Test SW-3f$_2$. Site test (left), FEM simulation (right) node no. 4 (fig. 4.17).

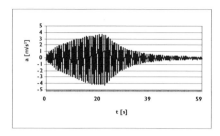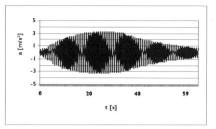

Figure 4.22 Footbridge in Sromowce, Test HC-12f$_1$. Site test (left), FEM simulation (right).

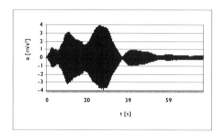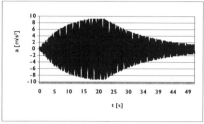

Figure 4.23 Footbridge in Sromowce, Test HC-12f$_2$. Site test (left), FEM simulation (right) node no. 4 (fig. 4.17).

4.4   DECK LATERAL VIBRATIONS UNDER VERTICAL EXCITATION AND VIBRATIONS OF STAYS

Using the presented above procedure for simulation of dynamic response of a pedestrian bridge, several other phenomena can be noticed. The most important are:

- lateral vibrations under vertical periodic load,
- vibration of stays,
- vibrations of secondary elements on the bridge. (street lamp mast, railing, etc.)

There are several possibilities to create the structure where lateral vibrations can be induced by pure vertical action. Two of them are presented below in fig. 4.24. The first is a cable-stayed bridge with unsymmetrical, inclined pylon stabilized by a backstay cable. As we consider that the backstay cable is a high strength steel element with a relatively small cross section, it is clear that under the vertical load on the deck the head of the pylon gets lateral movement which at the end creates a lateral vibration of the deck. The FEM time-step procedure can show this effect using dynamic function (16) or (18). The second example is a suspension bridge with suspension cables additionally curved in plan. A small asymmetry in cross section in live load creates the difference in forces in the suspension cables. As the suspension cable has a horizontal curve it is clear that an unsymmetrical load on the deck generates a horizontal movement. Under dynamic pedestrian vertical load this kind of structure can be excited

Figure 4.24 Examples of structures that can be exited to lateral vibrations under the pure vertical dynamic load.

to lateral vibrations. FEM time-step procedure can show this effect using dynamic function (16) or (18).

By FEM dynamic time-step procedure with the third order theory we can observe dynamic response of stays on the cable stayed bridge. However, this kind of analysis is important mainly in two cases:

- If the first natural frequency of stay is in the range of pedestrian pacing rate,
- If the first natural frequency of stay is closed to one of the first natural frequencies of the solid part of the structure.

As we are able to simulate vibrations of stays there is no restriction to study the problem of other secondary elements on the bridge (street lamps, railings, banners, etc.). What is important in time-step analysis is the realistic estimation of damping.

4.5   CONCLUSIONS

The numerical simulation of dynamic response of the footbridge should be the standard procedure in designing. The complexity of the problems is mainly related to couplings between varying loading and transient structure response. Therefore a good FEM structural model and proper understanding of loading is necessary to succeed.

REFERENCES

[1] Zoltowski, K., "Pedestrian Bridge, Load and Response", Proceedings of the Second International Conference Footbridges 2005, Venice, Italy, pp. 247–248, December 6–8, 2005.

[2] Clough, R.W., Penzien, J., "Dynamic of Structures". McGraw-Hill, 1975.

[3] Matsumoto, Y., Nishioka, T., Shiojiri H., Matsuzaki, K., "Dynamic design of Footbridges". IABSE Proceedings p. 17/78. 1978.

[4] Lees, A., vanrenterghem, J., De Clercq, D., "Understanding how arm swing enhances performance in the vertical jump". Journal of Biomechanics, No 37, pp. 1929–1940, 2004.

[5] SOFiSTiK FEM SYSTEM, www.sofistik.com

[6]   Biliszczuk, J., Hawryszków, P., Maury, A., Sułkowski, M., "Design of a cable-stayed footbridge with deck made of glued-laminated wood". Structural Engineering Conference. International Conference on Bridges, Dubrovnik, Croatia, pp. 261–268, May 21–24, 2006.

[7]   Biliszczuk, J., Hawryszkow, P., Maury, A., Sulkowski, M., "Cable-stayed bridge made of glued-laminated erected in Sromowce Nizne". Proceedings of the Third International Conference Footbridges 2008, Porto, Portugal, July 2–4, 2008.

[8]   Zoltowski, K., "Pedestrian On Footbridges, Vertical Loads and Response.", Proceedings of the Third International Conference Footbridges 2008, Porto, Portugal, pp. 283–284, July 2–4, 2008.

Chapter 5

The role of dynamic testing in design, construction and long-term monitoring of lively footbridges

Álvaro Cunha, Elsa Caetano, Carlos Moutinho &
Filipe Magalhães
University of Porto, FEUP, Porto, Portugal

SUMMARY

This work aims to provide to footbridge designers a correct understanding of the usefulness of several types of dynamic testing tools that can support the design, construction, control and continuous monitoring of lively footbridges. Following the content of SYNPEX Guidelines, two different levels of assessment for the evaluation of dynamic properties of footbridges are considered and described in detail, based on two case studies: FEUP stress-ribbon footbridge, at Porto, and Pedro e Inês pedestrian bridge, at Coimbra. At last, the interest of long-term dynamic monitoring of lively footbridges requiring vibration control devices is stressed and illustrated.

Keywords: footbridge; dynamic; instrumentation; testing; modal identification; lateral vibration; response; monitoring.

5.1   INTRODUCTION

Although a comprehensive knowledge of materials and loads and a significant modelling capacity provide a high degree of understanding of the structural behaviour at the current state-of-art, numerous uncertainties remain present at the design stage of Civil Engineering structures. As a consequence, the corresponding dynamic properties and behaviour can only be fully assessed after construction. This fact has special importance in the context of pedestrian bridges, considering the narrow band of frequency excitation that frequently includes important bridge frequencies, and the typically low damping ratios of modern footbridges.

In this context, this work aims to provide to footbridge designers a correct understanding of the usefulness of several types of dynamic testing tools that can support the design, control and continuous monitoring of lively footbridges.

5.2   EVALUATION OF DYNAMIC PROPERTIES OF FOOTBRIDGES

5.2.1   Levels of assessment

Experimental characterisation of the dynamic behaviour of a lively footbridge is an important component of the project. According to the SYNPEX Guidelines [1], it can be developed based on two different levels of complexity:

- Level 1 – Identification of structural parameters, with the purpose of calibrating numerical models and eventually tuning control devices. Natural frequencies, vibration modes and damping ratios are the parameters of interest;
- Level 2 – Measurement of the bridge dynamic response under human excitation for assessment of comfort criteria and/or correlation with the simulated response.

The adoption of one of the above mentioned strategies depends on the characteristics of the structure and on the objectives of the study.

Level 2 tests can be characterised as standard tests that should be developed at the end of construction of any potentially lively footbridge, providing important information for design and verification purposes. Based on the results of these tests, the bridge owner may decide whether to implement control measures or not. It should be noted that the use of experimental tests to check the comfort class of a specific footbridge requires the performance of measurements covering all vibration phenomena considered in the development of design load models and involves the obtainment of characteristic values of the response.

Level 1 tests are required when it is clear that the dynamic behaviour of the footbridge is beyond acceptability limits and control measures are necessary. The appropriate design of control devices requires an accurate knowledge of structural parameters, namely natural frequencies and vibration modes.

Next sections summarise general guidelines for testing and data analysis of footbridges.

5.3   INSTRUMENTATION

5.3.1   Response devices

Given that acceptability limits for pedestrian comfort are generally defined in terms of acceleration, the usual measured response quantity is acceleration.

Accelerometers are sensors that produce electrical signals proportional to the acceleration in a particular frequency band, and can be based on different working principles. Three main categories can be employed in Civil Engineering measurements: (i) piezoelectric, (ii) piezoresistive and capacitive, and (iii) force-balanced.

Compared with the other types, piezoelectric accelerometers have several advantages, such as: not requiring an external power source; being rugged and stable in the long term, and relatively insensitive to the temperature; being linear over a wide frequency range. A serious inconvenience exists in applications involving very flexible structures, which is the limitation for measurement in the low frequency range. Piezoelectric accelerometers are not in effect capable of a true DC response, given that the piezoelectric elements only produce charge when actuated by dynamic loads. Many piezoelectric accelerometers only provide linear response for frequencies higher than 1 Hz, although some manufacturers produce accelerometers that operate for very low frequencies.

Both piezoresistive and capacitive, and force-balanced accelerometers are passive transducers, which require external power supply, normally an external 5 V DC–15 V DC excitation. These accelerometers operate however in the low frequency range, i.e., from DC to approximately 50–200 Hz, therefore being adequate for almost all types of measurements in Civil Engineering structures.

For most pedestrian bridges the frequency range of interest is 0.5–20 Hz. Accordingly, common specifications for accelerometers are:

- Frequency range (with 5% linearity): 0.1–50 Hz;
- Minimum sensitivity: 10 mV/g;
- Range: ±0.5 g.

5.3.2 Force and input sensor devices

The identification of natural frequencies, vibration modes and damping coefficients on a structure can be performed through forced, free or ambient vibration tests.

Forced vibration tests are the basis of the traditional modal analysis techniques and provide the most precise results, given that they rely on controlled inputs and outputs. This is particularly relevant for damping coefficient estimates, whose quality is in a great extent affected by measurement uncertainties.

Free vibration tests consist in the recording of the structural response associated with the sudden release of a tensioned cable or other device that originates an initial deviation from the equilibrium position of the structure. These tests are relatively inexpensive when conducted at the end of construction of the footbridge and provide accurate estimates of damping ratios of the excited modes. They constitute an alternative to forced vibration tests.

Ambient vibration tests employ the current ambient loads on the structure as input loads, assuming that the frequency content of these is approximately constant in the frequency range of interest. Although this hypothesis is not necessarily valid, very good estimates of natural frequencies and modal shapes can be obtained at the current state-of-art. In effect, not only the precision of sensors is currently so high, that the structural response can be measured for very small levels of vibration, but also powerful data processing techniques are available [2,3,4] that can be employed to identify modal parameters. The use of these techniques provides however a significant dispersion of modal damping ratios estimates.

Concentrating on forced vibration tests of pedestrian bridges, possible input devices are an impact hammer or a vibrator.

5.3.2.1 Impact hammer

Impact hammer excitation is the most well-known and simple form of providing a controlled input to a Mechanical Engineering structure or component. For Civil Engineering applications, the same technique can be employed, provided that the impact hammer has adequate characteristics. For these particular structures, one solution available in the market is the hammer represented in Figure 5.1 (left), weighting about 55 N, whose tip is instrumented with a piezoelectric force sensor, having a sensitivity of 1 V/230 N and a dynamic range of 22.0 kN. The hammer operates in the range 0–500 Hz. Given that pedestrian bridges are normally flexible and relatively small, the impact hammer meets for this type of structures one of the most interesting applications. It is noticed however that the energy input in the very low frequencies is very small, meaning that mode shapes of very low natural frequency are not possibly mobilised into a measurable level.

5.3.2.2 Vibrator

Vibrators employed in Civil Engineering applications can be of three different types: electromagnetic, hydraulic and mechanical. The shaker represented in Figure 5.1 (right) is one of the solutions available in the market, and weights around 800 N, operating in the range 0–200 Hz, and delivering a maximum force of 445 N for frequencies greater than 0.1 Hz. This device is configurable for excitation both in horizontal or vertical directions and is driven by means of a signal generator, which feeds the shaker amplifier. Typical generated signals for tests are sinusoidal or random. The measurement of applied force is possible through load cells installed between the shaker and the structure. Given the limitations in the amplitude of generated load, electrodynamic shakers can only be used for excitation of small and medium size structures. On the contrary, both hydraulic and mechanical shakers can be employed for excitation of large structures. Mechanical shakers based on the rotation of eccentric

Figure 5.1 Impact hammer (left) and electrodynamic shaker (right) for Civil Engineering applications.

masses apply a sinusoidal excitation in a varying frequency range. These devices are seldom used at the current state of art, given the significant requirements for the setup and operation.

5.3.2.3   Input measurement devices

One important topic in the testing of pedestrian bridges is the measure of input loads induced by pedestrians, both when walking alone or in groups.

The walking of a single pedestrian can be indirectly assessed through response measurement, provided that the dynamic characteristics of the walking platform are fully known. The direct assessment of the concentrated load applied by a pedestrian can be made through instrumentation of the walking platform with force plates. For a walking group, one important measure is the degree of synchronisation of pedestrians, which can be assessed by means of video recording and image processing. Former work developed by Fujino *et al* [5] has shown that the trajectory of pedestrians can be measured through measurement of the motion of pedestrians head and shoulders.

5.4   RESPONSE MEASUREMENTS

The performance of **Level 2** tests should consider the following items:

  i   Identification of critical natural frequencies and damping ratios;
 ii   Measurement of response induced by one pedestrian;
iii   Measurement of the response induced by a small group of pedestrians;
 iv   Measurement of the response induced by a continuous flow of pedestrians.

The verification of acceptability limits of vibration for a particular pedestrian bridge should be based on the results of these tests, considering the specific use of the bridge.

5.4.1   Measurements of ambient response for identification of critical natural frequencies

Tests should preferably be conducted on the bridge closed to pedestrian traffic, provided the transducers sensitivity is sufficiently high to capture ambient vibration response (typical acceleration peak amplitudes of the order of 2–5 mg).

Assuming that a preliminary dynamic analysis of the bridge has been conducted, providing an estimation of natural frequencies and vibration modes, the instrumented sections will correspond to the sections of maximum estimated modal response of the estimated critical frequencies.

In case that only one accelerometer is used for response measurement, the following procedure can be employed: for each measurement section, the sensor is mounted and the ambient response is collected, based on two test series. One of the series is collected, if possible, with the bridge closed to pedestrians, subjected to ambient loads, in order to eliminate the frequency content associated with pedestrian excitation. That procedure allows for an identification of the critical natural frequencies for vertical and/or lateral vibrations. The second series should be collected under the current pedestrian excitation and provides a better characterisation of bridge frequencies, as well as a measure of the intensity of vibrations under current use.

The choice of sampling rate and processing parameters should respect the following points:

- Assuming the frequencies of interest lie in the range 0.1–20 Hz, a sampling frequency of 50 Hz to 100 Hz should be selected. The acquisition equipment should include analogue filters in order to avoid aliasing errors, otherwise higher sampling rates may be required;
- Designating by f_{low} the expected lowest natural frequency of the bridge, the collected time series should have a minimum duration given by the formula

$$(A/f_{low}) [n - (n\text{-}1) \text{ overl}] \text{ (s)} \hspace{4cm} (1)$$

where A is a constant, with a value of 30 to 40, n is the number of records that will be employed in the obtainment of an average power spectral density (PSD) estimate of the response, and overl represents the rate of overlap used for that estimate. Current values of n are 8–10, and a common rate of overlap is 50%;
- Considering as an example a structure with a lowest natural frequency of 0.5 Hz, the averaging over a number n of records of 10, and an overlap rate of 50%, the minimum duration of the collected time series should be 330–440 s. So a total number of 33000 to 44000 samples should be collected at a sampling frequency of 100 Hz, leading to average power spectra with frequency resolution of 0.017 Hz to 0.0125 Hz;
- The collected time series should be processed in order to obtain an average PSD estimate. One procedure to form this PSD is as follows: Divide the collected series into n records, considering the defined overlapping rate; remove trend for each record; apply time window (Hanning window, for example) in correspondence; evaluate normalised PSD of each record; average the set of raw PSDs;
- The analysis of PSD estimates collected at one or various sections allows for a former identification of the prototype natural frequencies;
- The peak response of the series collected under current pedestrian walking should be retained for comparison with acceptability limits.

Example

FEUP footbridge is a very slender stress-ribbon concrete slab embedding four prestressed cables and it takes a catenary shape over the two spans, with a circular curve over the intermediate support. The two spans 28 m and 30 m long and the 2 m rise from the abutments to the intermediate pier were the starting points for the definition of the bridge structural geometry. The constant cross-section is approximately rectangular with external design dimensions of $3.80 \text{ m} \times 0.15 \text{ m}$. The intermediate support is made of four steel pipes forming an inverted pyramid hinged at the base, with horizontal resistance provided only by the prestressed cables.

Figure 5.2 shows a lateral view of this lively footbridge, as well as two average power spectra of vertical accelerations measured during 6 minutes at two points located at about 1/3rd of each span. The choice of these measurement points stemmed

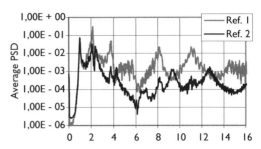

Figure 5.2 Lateral view of FEUP stress-ribbon footbridge (left); Average PSD spectra of the ambient response at two points.

from the fact that a previous numerical modelling indicated that the most relevant modes of vibration of the bridge have significant modal components at those two points. Inspection of those spectra allowed an accurate preliminary identification of the bridge critical frequencies.

5.4.2   Raw measurement of damping ratios associated with critical natural frequencies

Raw estimates of damping ratios associated with critical natural frequencies can be obtained from a simple free vibration test in which a pedestrian jumps/bends knees/bounces on a fixed location at a particular frequency, trying to induce resonant response of the bridge for the corresponding vibration mode. After a few cycles of excitation, the pedestrian action is suddenly interrupted and the free vibration response is registered. The application of a single degree of freedom identification algorithm to the free decay response (eventually band-pass filtered, whenever close modes or noise are present) allows for a raw estimation of damping coefficient by segments of the time series. This process should be repeated a number of times, in order to provide average estimates of damping coefficient as a function of amplitude of oscillation. A plot of damping coefficient versus amplitude of oscillation can be made, where the amplitude of oscillation is taken as the average peak amplitude of oscillation within the analysed series segment.

Example
Preliminary dynamic tests were also performed at FEUP stress-ribbon footbridge in order to identify modal damping factors associated to the most relevant mode shapes in terms of human induced vibrations. The procedure adopted consisted of exciting the structure with a frequency close to the corresponding natural frequency, using a pedestrian skipping in a fixed position. After some cycles of oscillation, the structure falls in resonance. By sudden interruption of the motion, the free vibration response could be recorded, providing information to the application of the logarithmic decrement method. Figure 5.3 shows the measured response concerning the two first modes of vibration, at frequencies of about 1 Hz and 2 Hz. The measured damping factors were of 1.7% and 2.6% for

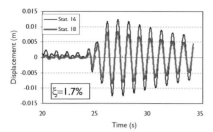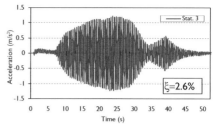

Figure 5.3 Free vibration response: Skipping at 1 Hz at 1/3rd (stat. 3) and ½ (stat. 5) of one span (left); skipping at 2 Hz (stat. 3) (right).

each one of these modes, respectively. It's worth noting, however, that this damping increases with the amplitude of oscillation. In fact, by introducing a high level of oscillation (4 m/s²) at 2 Hz, it was possible to identify a damping factor of 4.8%.

5.4.3 Measurement of the response induced by one pedestrian

The tests described above provide an update of the expected critical natural frequencies. The response is then measured at the relevant sections (maximum modal displacement section for each critical frequency), considering the motion of a single pedestrian over the bridge. Several types of motion should be explored, as a function of the frequencies of interest: walking, for critical natural frequencies below 2.5 Hz; walking and running, for critical natural frequencies between 2 Hz and 3 Hz; running, for natural frequencies above 3 Hz. Given the random characteristics of excitation, a number of realisations should be performed for each combination frequency/motion. A reference number is 5. In order to facilitate the definition of walking rate, a metronome should be used. The maximum acceleration and dynamic displacement of the bridge should be registered for each collected series, and the peak response induced by one pedestrian can be taken as the maximum of the peak responses registered for the various realisations. The weight of the pedestrian should be retained. Whenever the bridge exhibits non-symmetric slope, the response should be recorded for motion of the pedestrian in the declining sense.

5.4.4 Measurement of the response induced by a group of pedestrians

The response should be measured in two conditions: walking/running of group under current use, and walking/running of the group with the fit of inducing high response (vandalism). Looking in the literature, it can be noticed that the number of pedestrians used in group tests varies in the range 10–20 persons. For reference, it can be considered whenever possible that the group is formed by 10 pedestrians, if the deck width is not greater than 2.5 m, and 15 pedestrians, for larger widths.

The response should be measured based on the considerations made in the previous Section for the crossing of one pedestrian, ie., for each motion type/frequency combination, 5 realisations of one crossing of the bridge in declining sense (for non-symmetric slope) should be collected, at a sampling frequency of 50 Hz–100 Hz. The weight of the group members should be retained, and the group response should be the highest of the peak responses recorded. The series associated with the synchronised group should be collected making use of a metronome, in order to facilitate synchronisation in a particular frequency.

Given that it is expected that the presence of people on the deck might originate an increase of damping ratios and that, for high amplitudes of vibration these ratios increase, it is suggested that measurements are made of the free vibration response after resonant excitation of the bridge by the group, jumping on a fixed position.

Example

In addition to dynamic tests with small groups of pedestrians on the FEUP stress-ribbon footbridge, walking or running with step frequencies close to the bridge natural frequencies, measurements were made of the response induced by small groups of students jumping in synchronised form so as to induce resonance. Figure 5.4 shows an example of such a test for the excitation frequency associated with the second vibration mode, at 2.07 Hz. It can be observed that within less than 5s a group of 18 people induced a maximum acceleration of 6.6 m/s^2 at the third of the largest span. The corresponding displacement at that location was 3.4 cm, and the identified damping ratio ξ was 4.8%, whereas vibration tests at much reduced amplitude of oscillation led to a ξ value of 1.5%. Figure 5.4 also shows the variation of the maximum acceleration achieved as function of the number of pedestrians jumping together. Inspection of this plot shows the difficulty to exceed 1 g of maximum acceleration, owing to the increase of damping with the level of oscillation and the higher difficulty of

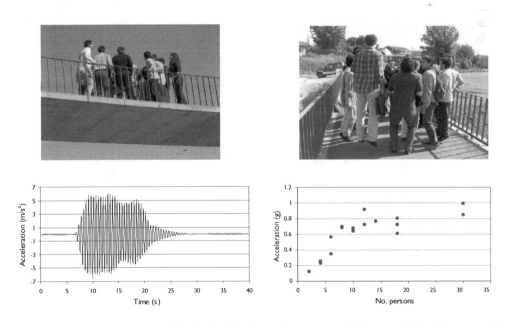

Figure 5.4 "Vandal" excitation of footbridge (top), acceleration record (bottom, left) and increase of maximum acceleration with the number of pedestrians jumping together (bottom, right).

synchronization with larger groups. These results show the enormous easiness of excitation of the footbridge at resonance and raise the question of the amount of damage that can intentionally be produced by a group of pedestrians.

5.4.5   Measurement of the response induced by a continuous flow of pedestrians

The measurement of the response induced by a continuous flow of pedestrians is a practice of interest for the characterisation of the footbridge response under different usage conditions and should be considered in particular for footbridges that clearly exhibit a lively behaviour, namely a trend for synchronisation effects. The measurement procedures are identical to the ones adopted for single pedestrian and group tests described in Sections 4.3 and 4.4.

Example
FEUP stress-ribbon footbridge exhibits several natural frequencies in the range 2–4 Hz, associated to vertical bending modes, that can be easily excited by pedestrians walking or running. With the purpose of characterizing the levels of acceleration induced by continuous flow of pedestrians, different series of tests were conducted with a large group

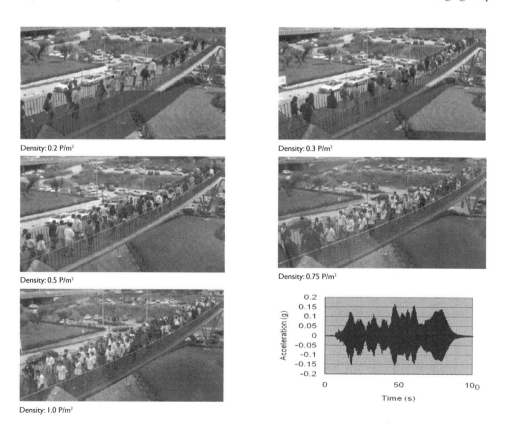

Figure 5.5 Crossing of FEUP footbridge with increasing human densities; Acceleration record at 1/3rd span.

of about 120 students. Different distributions of the group along the bridge were tested and two situations of flow were considered: normal walking and synchronised walking, by use of a metronome. Figure 5.5 shows several images of the group with different human densities and a typical acceleration record. It could be observed that the maximum pedestrian density on the bridge was about 1person/m², and the maximum vertical acceleration didn't exceed 2 m/s² (Figure 5.6, left), in a situation of synchronisation. For that case, maximum lateral acceleration of 0.5 m/s² was attained. Although these values are rather high and classified by the students as being in the limit from disturbing to strongly disturbing (Figure 5.6, right), they are much inferior to the accelerations produced intentionally by a small group of pedestrians, as shown in the previous section.

Pedro e Inês footbridge, over Mondego river at Coimbra, is also a lively bridge where very significant lateral vibrations might occur owing to a lock-in effect associated to the first lateral natural frequency of about 0.83 Hz, which required the implementation of a set of tuned mass dampers.

The bridge is a slender structure with a length of 275 m and a width of 4 m, except in the central square with dimensions of 8 m × 8 m (Figure 5.7). The metallic arch spans 110 m and rises 9 m and has a rectangular box cross-section with 1.35 m × 1.80 m. The deck has a L-shaped box cross-section with its top flange formed by a composite steel-concrete slab 0.11 m thick (Figure 5.8). In the central part of the bridge, each L-shaped box cross-section and corresponding arch "meet" to form a rectangular box cross-section 8 m × 0.90 m. In the lateral spans, arch and deck generate a rectangular box cross-section 4 m × 0.90 m.

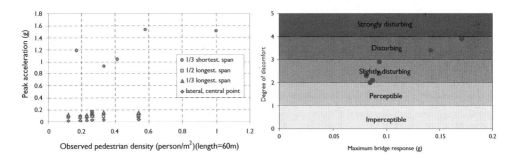

Figure 5.6 Peak vertical or lateral accelerations measured at different points of FEUP footbridge, as function of the observed pedestrian density (left); Degree of human discomfort expressed by the students as function of the bridge maximum response (g).

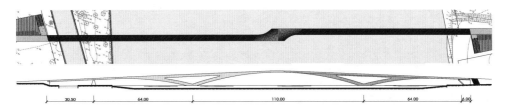

Figure 5.7 Plan and lateral views of Pedro e Inês footbridge.

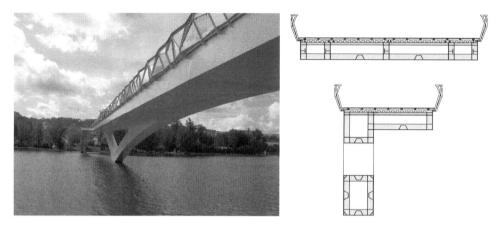

Figure 5.8 Cross section at the central square and at an intermediate position of the arch.

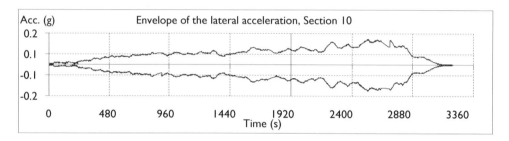

Figure 5.9 Envelope of lateral acceleration at mid-span.

Figure 5.10 Variation of the amplitude of lateral accelerations at mid-span with the number of pedestrians.

With the purpose of analysing experimentally the lateral behaviour of the bridge, a test was developed during which the bridge response was measured to the action of a continuous stream of pedestrians with a gradually increased number to a maximum of 145 students from the Universities of Porto and Coimbra. These students walked freely along the central part of the bridge. Figures 5.9 and 5.10 show the envelopes of the measured lateral accelerations at mid-span along the time and the variation of the maximum lateral acceleration with the number of pedestrians on the bridge, respectively. At that section, extreme values of acceleration of ±1.2 m/s², and of displacement of ±4 cm, were recorded when 145 pedestrians were walking on the bridge. Figure 5.10 shows that the increase of acceleration with the number of pedestrians on the bridge is not linear, but instead exhibits a "jump" precisely for values of the number of pedestrians close to 70, which is coherent with the estimate provided by the formula developed by Dallard *et al* [6] for the Millennium Bridge.

5.5   MODAL IDENTIFICATION TESTS

The identification of modal parameters, i.e., natural frequencies, vibration modes and damping coefficients, is performed through the above designated Level 1 tests. Accordingly, a conventional modal analysis technique can be applied, based on forced vibration tests or, alternatively, identification can be based on free or on ambient vibration tests. The basic parameters of the tests are established for the two cases in the following sections.

5.5.1   Forced vibration tests

The identification technique to apply depends on the type of excitation employed. A wide band excitation defined in the frequency range of interest is most convenient in terms of providing a quicker estimation of parameters. However, there is a risk that the input energy associated with the low natural frequencies is very small and so the signal-to-noise ratio may be very low.

The general procedure for the identification of mode shapes is presented, considering particular aspects associated with each type of excitation.

5.5.1.1   *Hammer excitation*

Even with the softest tips, hammer excitation produces a short duration pulse (typically 10 ms, on a concrete surface), whose frequency content is defined in a wide range, such as DC–200 Hz. Although analogue filters may be incorporated in the conditioning or acquisition equipment, the spectral content of the input can only be accurately defined if the time description is accurate. Assuming this pulse is represented by a half sinusoid, three points should be used to describe accurately this curve, with a minimum spacing of 5 ms. Hence, a minimum sampling frequency of 200 Hz should be employed, even though the frequency content of interest lies in the range 0.1 Hz–20 Hz.

Another aspect to retain is that, given that the input force is applied manually, some differences in the quality of the signal applied may occur. In particular, it is important for the operator to avoid double hits in each recorded time series, which significantly affect the quality of frequency response estimates.

Referring to the length of each recorded time series, it should be defined, if possible, in such a way that the structural response to hammer impulse vanishes within the collected series. In that case, time windowing is not necessary, therefore increasing the quality of damping estimates. A reference maximum duration of the series is 20.48 s, corresponding to a number of 4096 points sampled at 200 Hz. This corresponds to obtaining spectral estimates with a frequency resolution of 0.04 Hz, which is manifestly insufficient to characterise mode shapes at very low frequencies. Hammer excitation should not in effect be used for the characterisation of those modes. It should be noted that, even though longer records can be collected, the last part of the signal may contain only ambient vibration response and therefore does not provide an input correlated signal.

Assuming the sampling frequency and duration of records are defined, one procedure for the obtainment of a set of frequency response function estimates is as follows:

i   Selection of a section along the deck where to apply the hits. This section should be chosen considering preliminary numerically calculated mode shapes, in such a way that a minimum number of modal nodes are close. More than one section may have to be defined, depending on the configuration of mode shapes of interest;

ii For each input section R_j, and depending on the number of available accelerometers, install successively the accelerometer(s) on the measurement sections. For each (set of) instrumented section(s), using the sampling parameters above defined, collect the response to the impulse hammer applied at R_j, as well as the input signal at the force sensor. For each set-up, a total of 5 to 10 time sets of series are recorded;

iii Remove trend to all response time series. Obtain a spectral description of the input and response, through estimation of auto-power spectra $\tilde{S}_{ii}(f) =$ and $\tilde{S}_{jj}(f)$. Estimate the cross-spectrum $\tilde{S}_{ij}(f)$ relating the response at each measurement section R_i, with the input applied at section R_j. Average the set of auto and cross power spectra, for the set of 5 to 10 series collected at each location

$$S_{jj}(f) = E\left[\tilde{S}_{jj}(f)\right]$$
$$S_{ij}(f) = E\left[\tilde{S}_{ij}(f)\right]$$

(2)

Estimate frequency response functions $H_{ij}(f)$, based on estimator H_2.

$$H_{ij}(f) = \frac{S_{ij}(f)}{S_{ii}(f)}$$

(3)

and coherence $\gamma^2(f)$, defined as

$$\gamma^2(f) = \frac{\left|S_{ij}(f)\right|^2}{S_{ii}(f) \cdot S_{jj}(f)}$$

(4)

The functions $H_{ij}(f)$ are intrinsic of the system and form the basis for application of a System Identification algorithm (in the frequency domain) to extract natural frequencies f_k, vibration modes φ_k and associated damping coefficients ξ_k, while $\gamma^2(f)$ provides a measure of the correlation between the measured input and response signals.

Considering a viscous damping model and response measurements expressed in terms of accelerations, the frequency response functions $H_{ij}(f)$ relate to the modal components of mode k, $(\varphi_i)_k$ and $(\varphi_j)_k$ at sections R$_i$ and R$_j$, respectively, through

$$H_{ij}(f) = \frac{-f^2 \cdot (\varphi_i)_k \cdot (\varphi_j)_k}{(f_k^2 - f^2) + i(2\xi_k f_k f)} \qquad (5)$$

Example
Conventional modal analysis tests were performed at FEUP stress-ribbon footbridge using an impact hammer, a set of four piezoelectric accelerometers and an eight-channel Fourier analyzer. The impact hammer has a force sensor at the tip with a sensitivity of 1V/0.23 kN, which allows the measurement of the applied force. The piezoelectric accelerometers have a sensitivity of 1V/g. These devices were connected to signal conditioners for further amplification of the signal. The impulsive loads

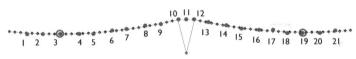

Figure 5.11 Application of hammer impulse (left); Measurement points in modal identification tests (right).

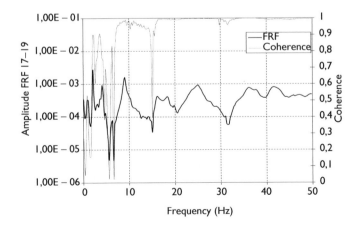

Figure 5.12 FRF and coherence relating force applied at node 19 with acceleration measured at node 17.

were always applied (Figure 5.11) at a fixed point at each span (nodes 3 or 19), and the corresponding response was measured at four different points in each setup. The evaluation of frequency response functions (FRFs) relating the input force and the output accelerations was performed in the frequency range 0–200 Hz, considering an individual time of acquisition of 16s and the average over 8 spectral estimates, which led to a frequency resolution of 0.0625 Hz.

Figure 5.12 shows one the FRFs obtained, relating the input force and output acceleration at nodes 19 and 17, respectively, as well as the respective coherence function. Inspection of all the FRFs obtained shows that their peaks allow to easily identify the fundamental natural frequencies of the footbridge (0.979, 2.050, 2.446, 3.629, 4.104 and 5.374 Hz). However, the relatively short time of each acquisition (16s), associated to the application of each impulsive load, led to a relatively low frequency resolution (0.0625 Hz) of the spectral estimates, whose peaks became wider. Accordingly, the separation of close modes and the accurate identification of modal damping factors (clearly overestimated) became impossible.

On the other hand, despite the very high values of the coherence functions in the frequency range 10–200 Hz, their values fall drastically in the frequency range (0–10 Hz), evidencing the difficulty of the hammer to excite significantly the most relevant modes of vibration in that range, with regard to the level of noise induced by ambient factors, like wind.

These aspects precluded the obtainment of reliable estimates of modal components.

5.5.1.2   Vibrator excitation, wide-band

Wide-band excitation induced by hydraulic or electrodynamic vibrators can be of transient or continuous type. Transient signals, like burst random, are treated in a way similar to those produced by hammer excitation. Continuous signals require time windowing applied to each time segment of the series, in order to reduce leakage effects. Moreover, since time windowing reduces the contribution of the edge samples, it is frequent to overlap time segments. A common procedure consists in the application of Hanning windows to the input and response time segments, combined with an overlapping rate of 50%. This allows a considerable reduction of the duration of the time series collected at each pair of input-output sections. Common wide-band generated signals are random or chirp-sine.

Example
FEUP footbridge was also studied performing a conventional modal analysis test based on an electrodynamic shaker controlled by a Fourier analyzer (Figure 5.13).

In a first instance, the shaker was placed in the longer span (at nodes 3 or 5), applying a random load in the frequency range 0–6.25 Hz, the corresponding structural response being measured by piezoelectric accelerometers at four different points in each setup. The evaluation of frequency response functions (FRFs) relating the input force and the output accelerations was performed, in this case, in the frequency range 0–6.25 Hz, considering an individual time of acquisition of 128s, an average over 8 spectral estimates (with 50% overlapping), and applying a Hanning window, which led to a higher frequency resolution (0.01563 Hz).

As some of the modes of vibration only involve the motion of one of the spans, the shaker was subsequently placed in the shorter span (at nodes 19 or 17).

(a)   (b)

Figure 5.13 (a) Application of random shaker load; (b) Detail of the shaker and load cells.

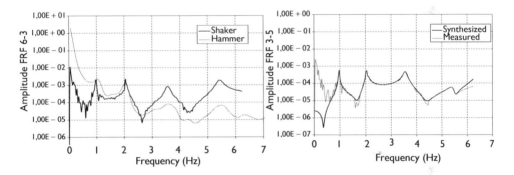

Figure 5.14 Comparison of FRFs (shaker vs hammer) (left); Measured vs synthesized FRFs (right).

Figure 5.14 (left) shows a comparison of two estimates of a FRF, obtained through the application of the shaker and the hammer. This plot shows two interesting aspects: (i) the much better definition of the peaks of the FRFs allowed by the higher frequency resolution of the spectral estimates associated to the shaker; (ii) the higher capacity of the shaker to excite the fundamental modes of vibration of the footbridge, with frequencies below 10 Hz.

The identification of the modal parameters of the bridge was done by applying a multi-degree-of-freedom identification algorithm in the frequency domain (RFP method) [7] to the FRFs obtained. Figure 5.14 (right) shows, for instance, a comparison between amplitudes of one of the measured FRFs and the corresponding synthesized FRF, on the basis of the identified modal parameters, showing the good agreement achieved.

5.5.1.3 Vibrator excitation, sinusoidal tests

The performance of sinusoidal tests provides the best results, as long as the vibrator has sufficient power to induce the vibration modes of interest. This point is critical for very low natural frequencies, even though pedestrian bridges are very flexible.

The procedure for construction of frequency response functions and identification of vibration modes and damping coefficients comprehends a preliminary collection of ambient response, which provides an approximation of natural frequencies. Once the vicinity of each natural frequency of interest has been identified, a sinusoidal test is developed that consists in the construction of parts of the frequency response function, point by point, each point corresponding to the pair frequency of excitation, frequency content of the measured response at each measurement section. The following points should be considered:

i   Although it is desirable to measure the applied force, that is not always possible, particularly if an eccentric mass shaker is employed. The force applied by such type of shakers can however be estimated with a certain precision;

ii   The precise identification of the natural frequency of the structure is made by application of a sinusoidal excitation and recording of the response at one particular location where the estimated mode shape has a significant component. For each excitation frequency a time series of the response at a particular location can be extracted, with a short duration, corresponding for example to 512 samples. Assuming the induced signal is a perfect sinusoid, the amplitude and phase of the response can be extracted by single degree-of-freedom time domain data fit. The frequency response function dot is obtained by the ratio to the input excitation amplitude measured or estimated;

iii   Although very short time series are required, it is necessary that the shaker operates for each frequency for a period of at least one minute, in order to guaranty that stabilisation of the response has been achieved;

iv   Once the natural frequency has been identified, the vibrator is tuned to that frequency and one accelerometer, or a set of accelerometers are successively mounted at each measurement location to collect a small time series of response. When a force sensor is not employed, it is necessary to install an accelerometer close by the vibrator, which remains fixed. Simultaneous records of response at two locations are then collected, for an evaluation of the relative phase and amplitude to the reference section. The set of amplitudes and phase ratios to the reference point constitute mode shape components;

v   The best quality of damping estimates is obtained with sinusoidal tests. Damping estimates are derived from the analysis of the measured free vibration response after sudden interruption of sinusoidal excitation at resonance. Provided that no close modes are present, a single degree-of-freedom algorithm is sufficient to identify the damping ratio. Given that this ratio depends on the amplitude of response, the free vibration response should be analysed by segments of the response record.

Example

The electrodynamic shaker was also used at FEUP footbridge to induce modal resonance, by applying sinusoidal excitations with frequencies in correspondence with some of the previously identified natural frequencies. By stopping suddenly that excitation, it was possible to measure the corresponding free vibration response (Figure 5.15) at several points, and extract some accurate estimates of modal damping factors.

At Pedro e Inês footbridge, forced vibration tests using sinusoidal excitations were essential for the evaluation of the real efficiency of passive control devices introduced at mid-span to attenuate lateral vibrations and avoid serious lock-in phenomena. The option of the designer was to install 6 TMD units (Fig. 5.16, left), with a total mass of 14970 kg, corresponding to a mass ratio $\mu = m_T / M_H \times 100 = 7.3\%$, which should provide a minimum theoretical damping of 7.8%. This option stems essentially from the knowledge of the sensitivity of the TMDs with regard to the achieved frequency tuning.

The intermittent and different activation of the various TMD units raised the question of the accuracy of the frequency tuning of the individual units and of the global efficiency of the ensemble, motivating the performance of a set of forced vibration tests for an accurate characterisation of the installed control system. These tests were developed using an exciter built at the Mechanical Engineering Department of FEUP, based on a hydraulic actuator that induced the horizontal sinusoidal displacement of a sliding mass of 1200 kg, enabling the application to the structure of a horizontal sinusoidal force with maximum amplitude of 1300 N. This mechanical device was installed at the central square of the bridge (Figure 5.16, right), applying swept-sine excitations in the frequency range 0.58 Hz–0.99 Hz with an increment of 0.01 Hz and with two different amplitudes of 200 N and 1300 N. For each step frequency 30 cycles of force were generated. Aiming the characterisation of the bridge behaviour with and without TMD, and also of the individual behaviour of each unit, the tests were

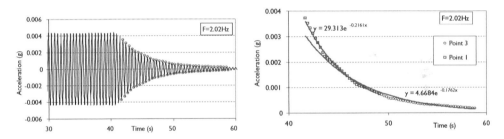

Figure 5.15 Free vibration modal response and fitting of the free decay response envelope.

Figure 5.16 Horizontal TMD installed at the midspan section, formed by 6 similar units (left); Mechanical exciter used in the forced vibration tests with swept-sine excitation (right).

performed in the following situations: (i) all TMD units blocked; (ii) all TMD units free; (iii) 1 TMD unit free and the other 5 blocked, for each one of the 6 tested units.

Figure 5.17 shows the amplitude of the frequency response functions (FRFs) obtained experimentally, relating the measured acceleration at midspan of the deck and at the different units of the TMD with the applied force, considering amplitudes of 200 N (Figure 5.17 (a) and 1300 N (Figure 5.17 (b).The force of 200 N is clearly not sufficiently high to activate the TMD, and so the obtained FRF is characterised by the frequency and damping of the non-controlled system, f = 0.83 Hz, ξ = 0.55%. The FRF obtained with the force of 1300 N corresponds to a totally different system, with clear influence of the TMD and revealing two modes of vibration with the following identified characteristics: f_1 = 0.79 Hz; ξ_1 = 6.5%; f_2 = 0.88 Hz; ξ_2 = 4.0%.

Considering all the TMD units released, the application of swept-sine excitations was repeated in the frequency range 0.58 Hz–0.99 Hz. Figure 5.18(a) presents a time record simultaneously captured at the deck level and at each one of the TMD units during the sinusoidal excitation at a frequency of 0.82 Hz with force amplitude of 1300 N. Figure 5.18 (b) illustrates the variation with frequency of the amplitude of

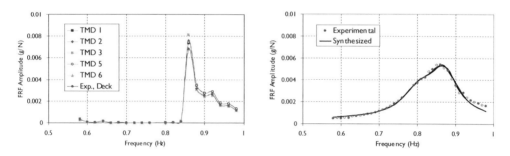

Figure 5.17 Comparison between FRFs with different applied force amplitudes: (a) deck and TMD for F = 00N; (b) deck for F = 1300N.

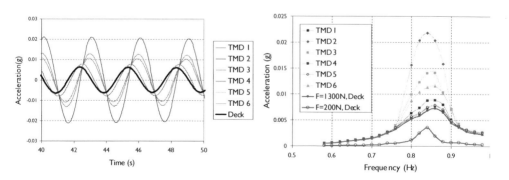

Figure 5.18 (a) Temporal evolution of lateral accelerations of deck and TMD units for a frequency excitation of 0.82 Hz; (b) Acceleration amplitude at the deck and TMD units for swept-sine excitation.

acceleration recorded at the deck and different TMD units. Inspection of these figures shows that the response of the several units presents some phase shift and distinct amplitude, and also that not all TMD units start their motion simultaneously, which evidence the different levels of damping associated to those units.

5.5.2   Ambient vibration tests

Ambient vibration tests are becoming an extremely attractive alternative for identification of modal parameters in Civil Engineering structures, given the limited required resources, and the high precision of currently available sensors.

As previously mentioned, the basic hypothesis for ambient vibration tests is that the input, i.e., the ambient excitation, can be idealised as a white-noise defined in a bandwidth corresponding to the frequency range of interest. This means that, within a certain frequency range, all mode shapes are excited at a constant amplitude. The recorded response is therefore an operational response, and the technique of constructing so-called frequency response functions, relating the responses at two measurement sections (transfer functions), leads to identification of operational deflection shapes, instead of modal shapes. Assuming that the frequencies of the system are well separated, and that the damping coefficients are low, a good approximation exists between operational deflection shapes and modal shapes. However, if frequencies are close, the operational deflection modes comprehend a non-negligible superposition of adjacent modes, therefore providing erroneous results. Although some possibilities exist for providing a separation of mode shapes, like separating bending and torsional response on a bridge by constructing two signals, the half-sum and half-difference of the edge deck measured response, some other alternatives are offered in terms of signal processing, that allow identification of modal components and damping coefficients. That is the case of the stochastic subspace identification methods, which are output-only parametric modal identification techniques that can be applied directly to acceleration time series or to the corresponding response covariance matrices [2]. These methods have been implemented in a toolbox for Matlab (Macec) [8]. Also commercially available is a software based on the stochastic subspace identification and frequency domain decomposition methods (Artemis) [9], as well as another one based on Polymax method [10], which are also powerful tools for modal identification.

Although damping estimates are provided by the more powerful algorithms, the precision in the estimates is limited and so results should be used with care.

The conventional technique for identification of operational deflection shapes requires the building of frequency transfer functions between outputs. This is done exactly as described in Section 5.1 for forced vibration tests with wide-band excitation.

Example

The final design of the TMDs of Pedro e Inês footbridge required the accurate identification of its modal properties after construction. That was based on the performance of an ambient vibration test using four seismographs, equipped with force-balance tri-axial accelerometers and 18bit analogue-digital converters, dully synchronized by external GPS units.

During the one and a half days of testing, the 20 sections indicated in Figure 5.19 were instrumented. Except for sections 9, 10 and 11, where accelerations were

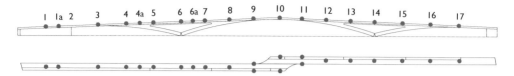

Figure 5.19 Instrumented sections: lateral and top view (reference points are indicated in blue).

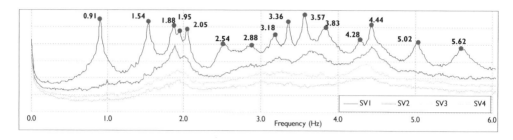

Figure 5.20 Averages of the first four normalized singular values of the spectra matrices.

measured at both sides of the deck (upstream and downstream), to better characterize torsional effects, vibrations were just recorded at the longitudinal axle of the deck.

Knowing that several vertical modes of vibration were of local nature, and aiming to identify as many modes as possible, three of the measurement units were used as references, permanently located at sections 1, 6 and 8, while the fourth one was successively placed at the remaining measurement points. Acceleration time series with 16-minute duration were recorded at 100 Hz sampling rate for each setup. The natural frequencies and modes of vibration were evaluated using the Frequency Domain Decomposition (FDD) and the Data driven Stochastic Subspace Identification (SSI-DATA) methods.

Figure 5.20 shows the averages of the first four normalized singular values associated with 19 spectra matrices. In this graphic it is possible to identify 15 natural frequencies in the analysed frequency interval (0–6 Hz). Detailed presentation of identified modes can be shown at ref. [11]. Figure 5.21 shows, for instance, the transversal (t) and vertical (v) components of the first identified modes of vibration, which are also compared with the corresponding calculated components. It is observed that this excellent correlation could be only achieved based on the development of an entirely new numerical model, using a very refined mesh of shell finite elements reproducing the bridge geometry. In this model all the openings in the deck were modelled and the transversal stiffening elements were included by equivalent beam elements, the stiffening constants of the springs at the foundations of the arches having been iteratively adjusted [12].

5.5.3 Free vibration tests

Considering that the sudden release of a tensioned cable is equivalent to the application of an impulse, the identification of modal parameters from a free vibration test

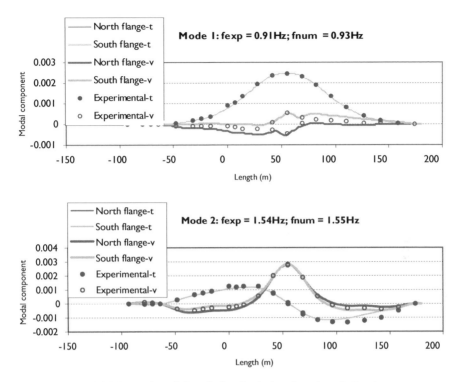

Figure 5.21 Examples of identified and calculated modes of vibration.

can follow the procedure described in Section 5.1.1, in which the frequency spectrum of the input is assumed constant for the range of analysis. Alternatively, the output-only identification algorithms of the type described in Section 5.2 can be applied. In any case, it is expected that higher quality modal estimates are obtained than those resulting from ambient vibration tests.

Example
The free vibration tests of Pedro e Inês footbridge were developed to achieve very accurate estimates of the damping ratios of the modes of vibration susceptible to be more significantly excited by pedestrians, as those values were determinant for the final design of the TMDs.

The impulsive loads were applied by the sudden release of a mass previously suspended from the structure. A mass of 4.5 ton was used for vertical impulses at sections 10 and 6 (Figure 5.19), while 3 ton were used for the other cases. Figure 5.22 illustrates some of the procedures and equipment used during these tests. The impulsive excitation was repeated at least two times in each case, in order to increase the accuracy of the estimates.

As the excitation of the fundamental lateral mode by pedestrians walking on the bridge could lead to lock-in phenomena, previously anticipated at design level, particular attention was dedicated to the accurate identification of the corresponding damping ratio. For this purpose, an auxiliary metallic structure with a pulley (Figure 5.22a)

(a)   (b)   (c)

Figure 5.22 Application of impulsive loads: (a) structure developed to apply the lateral impulse; (b) suspension of the mass at section 6; (c) cut of a cable hanging the mass from the bridge.

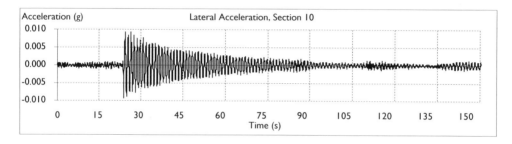

Figure 5.23 Lateral acceleration time series at section 10.

was used to allow the application of a horizontal impulsive force by sudden release of a mass of 3t (Figure 5.22c). Figure 5.23 shows a record of the lateral acceleration measured at mid-span during the release of that mass, which was clearly dominated by the frequency of the first lateral mode (f = 0.91 Hz, at that stage of construction). The release of the mass was also repeated three times, two of them on the first day under moderate wind (average wind velocity of about 8–10 m/s) and the third on the second day, with almost no wind. The exponential fitting of the envelope of the free vibration lateral response at section 10 led to the evaluation of the modal damping ratio of the fundamental lateral mode, which oscillated between 0.52% and 0.65%.

5.6 LONG-TERM DYNAMIC MONITORING

In case of lively footbridges where human comfort limits may be exceeded, and specially where lateral lock-in phenomena may occur or vibration control devices have been implemented, temporary or long-term dynamic monitoring can play a very important role, providing continuous information about the footbridge dynamic behaviour and sending alert messages if some threshold is overcome. Appropriate software must be then also developed and implemented for the acquisition, analysis, data processing, statistical characterization and results visualization, allowing the remote control through the Internet.

Example

With the purpose of permanently monitoring the dynamic response of Pedro e Inês footbridge during 12 months after construction and detect eventual episodes of excessive vibration, the bridge was instrumented with 6 uniaxial piezoelectric accelerometers installed in correspondence with the location of the TMDs, which were implemented at the antinodes of the critical vibration modes. Five of these accelerometers measure vertical accelerations, whereas another one measures lateral vibrations at mid-span (Figure 5.24). These sensors are installed inside the metallic deck and are wired to an acquisition system located inside one of the concrete abutments of the structure. By this means, the equipments are protected against some undesirable external environment like humidity and dust, and at same time the level of security against vandalism is increased.

The acquisition system used for the dynamic monitoring of the footbridge is composed by a signal conditioner, a digital computer incorporating an analogue-digital conversion board and a UPS system. The signal conditioner amplifies the signal from the accelerometers integrated in a ICP circuit and performs some basic signal processing like analogue filtering. The data acquisition is carried out by an A/D card fitted in a digital computer based on software developed with LabVIEW from National Instruments. The UPS improves the performance of the monitoring system by stabilizing the electrical power and covering some eventual gaps in power supply. The equipments are prepared to face prolonged loss of electrical power by the switching of some automatic shutdown mechanisms complemented by the automatic restart in the case of normalization of the situation. Beyond that, there is also a communication system responsible to transmit data to FEUP using an ADSL line. This communication system sends permanently to a computer located at FEUP the most recent data to be processed, which makes possible the virtually on-line monitoring of the structure.

The basic architecture of the whole monitoring system is schematically illustrated in Figure 5.25, which divides this system in three distinct modules corresponding to the different stages of the monitoring process. The first module involves the signal acquisition and conditioning at the Coimbra footbridge and the respective storage in a local database. It is important to combine these components because, in case of a communication failure, the data acquisition system keeps collecting signals to the local database. After the reestablishment of the communications, the data corresponding to that period is transmitted to FEUP in such way that there is no data loss during this anomaly.

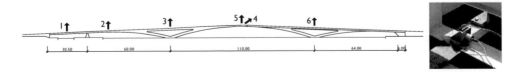

Figure 5.24 Location of the accelerometers (left); Accelerometer used to measure lateral accelerations at midspan (right).

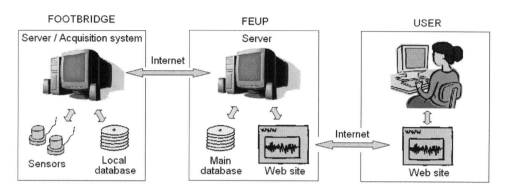

Figure 5.25 Schematic representation of the architecture of the monitoring system.

The second module of the monitoring system includes all the operations related with signal processing carried out at FEUP. After transmission from the footbridge site, the signals are organized in a main database which can be accessed at any time by some post-processing tools. In order to satisfy the main objective of this monitoring system, a web site was developed which allows the visualization of the time signals of the six accelerometers and subsequently the monitoring of the vibration levels of the structure (Figure 5.26). In addition, some spectral analysis routines are available, which involve the calculation of the Average Normalized Power Spectrum Densities (ANPSD), in order to evaluate the frequency content of the signals. The corresponding graphics also allow to clearly identify the most relevant natural frequencies of the structure.

In order to enable an automatic and users friendly procedure to observe and interpret the very significant amount of data collected along several months, two other toolkits were developed in LabView.

The first one is the Automatic Analysis Toolkit, which allows the systematic processing of all the date captured in each month along successive periods of 20 minutes. By analysing and processing each file corresponding to such measurement period, this toolkit enables: (i) The elimination of spurious spikes in the signal, based on a wavelet analysis technique; (ii) The detection and evaluation of peak acceleration values related with the lateral and vertical vibration of the footbridge; (iii) The evaluation of mean maximum values of acceleration in successive short periods of duration; (iv) The representation of all time series in the frequency domain, by application of FFTs; (v) The statistical characterization of all data captured during each day or each month; (vi) The automatic identification of the first natural frequencies, mode shapes and modal damping ratios, based on the Peak-Picking (PP), Enhanced Frequency Domain Decomposition (EFDD) or Covariance-driven Stochastic Subspace Identification (SSI-COV) methods; (vii) To plot waterfall diagrams to detect eventual variations of natural frequencies during one day or one month; (viii) To save all the analysis results, including data and plots, both in time and frequency domains, in a complex database in a server computer.

The second toolkit is the Result Viewer Toolkit, which enables the easy visualization of all results and plots by simply pressing buttons. This result viewer system can also be published in the web, allowing the access through the Internet.

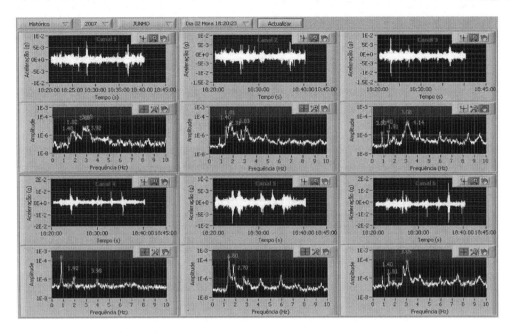

Figure 5.26 Web site to access the measured acceleration time series and corresponding Fourier spectra.

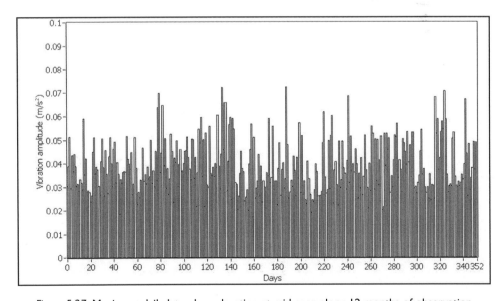

Figure 5.27 Maximum daily lateral acceleration at mid-span along 12 months of observation.

Figure 5.27 shows the variation of the maximum daily lateral acceleration at mid-span along 12 months of observation, showing that the limit level of lateral acceleration to avoid lock-in, assumed as 0.1 m/s^2, was never reached.

5.7   CONCLUSION

The examples used to illustrate the dynamic testing procedures suggested in the SYP-NEX Guidelines clearly show the important role that dynamic testing tools can presently play to support the design, construction and control of lively footbridges, allowing either a preliminary evaluation of the most relevant dynamic bridge properties and levels of human induced vibration, or the more accurate modal identification, based on ambient, free or forced vibration tests.

Last, but not least, long-term dynamic monitoring of lively footbridges, particularly in cases where vibration control devices are required to avoid lateral lock-in effects, is a solid basis to prove the real efficiency of such devices or to detect when human comfort limits are exceeded.

REFERENCES

[1] SYNPEX Guidelines, European Project on *Advanced Load Models for synchronous Pedestrian Excitation and optimised Design Guidelines for Steel Footbridges*, 2007.
[2] Peeters B., *System Identification and Damage Detection in Civil Engineering*, Ph.D. Thesis, Katholieke Universiteit Leuven, 2000.
[3] Brincker R., Zhang L. and Andersen P., "Modal identification from ambient responses using frequency domain decomposition", In Procceedings of IMAC-XVIII, International Modal Analysis Conference, pp.625–630, San Antonio, Texas, USA, 2000.
[4] Van Overschee P., De Moor B., *Subspace Identification for Linear Systems: Theory-Implementation-Applications*, Kluwer Academic Publishers, Dordrecht, The Netherlands, 1996.
[5] Fujino Y., Pacheco B., Nakamura S. and Warnitchai P., "Synchronization of Human Walking Observed during Lateral Vibration of a Congested Pedestrian Bridge", *Earthqauke Engineering and Structural Dynamics*, Vol.22, pp.741–758, 1993.
[6] Dallard, P. et al., The London Millennium Footbridge, *The Structural Engineer*, Vol. 79, No. 22, 2001.
[7] Han, M-C & Wicks, A.L., "On the application of Forsythe orthogonal polynomials for global modal parameter estimation", In Procceedings of IMAC-VII, International Modal Analysis Conference, pp.625–630, 1989.
[8] http://www.bwk.kuleuven.ac.be/bwm/macec/index.html
[9] http://www.svibs.com/
[10] http://www.lmsintl.com/testlab
[11] Magalhães, F., Cunha, A. & Caetano, E., "Dynamic testing of the new Coimbra footbridge before implementation of control devices", *International Modal Analysis Conference IMAC-XXV*, Orlando, Florida, SEM, 2007.
[12] Caetano, E., Cunha, A., *Estudo dinâmicos para avaliação das características dos TMDs da ponte pedonal e de ciclovia sobre o rio Mondego*, Relatório VIBEST, FEUP (in Portuguese, confidential), 2006.

Part II

Practical experience – case studies

After two large vibration episodes in two landmark footbridges, the Millennium Bridge in London, and the Solférino Bridge in Paris, the pursuit of construction of long span footbridges has put a lot of pressure on designers and researchers. Reaching acceptable levels of comfort implies in several cases the implementation of control devices. It is therefore of importance to learn from practice and to eventually include in the design the possibility to implement mechanical devices as a form to mitigate vibrations.

Wasoodev Hoorpah, Olivier Flamand and Xavier Cespedes focus on the studies and tests developed within the design and construction of the Simone de Beauvoir footbridge in Paris, discussing the implementation of dissipation devices and the need to monitor the footbridge behaviour throughout the first few years of operation.

Theodor Zoli shows that pedestrian induced vibrations are only one of the dynamic effects to consider in the design of long span footbridges, exemplifying with various US footbridge cases whose design has been governed by wind and earthquake actions.

One of the most complex problems to solve in the design of footbridges relates to the correct modelling of the action of a number of pedestrians and of their interaction with the footbridge vibrations. James Brownjohn, Stana Zivanovic and Aleksandar Pavic review the current knowledge on this topic and describe different approaches to model streams of pedestrians and crowds, presenting several cases of measured and simulated lateral and vertical responses.

Dissipation devices have been introduced in civil engineering applications for many years. However their design and implementation has scarcely been discussed in the literature.

Christian Meinhardt focuses on the design of tuned mass dampers (TMDs) and presents practical ways for the determination of the relevant dynamic behaviour and the in-situ assessment of the TMD effectiveness on the basis of constructed footbridges.

Philippe Duflot and Doug Taylor provide an overview of fluid damping technology and discuss design requirements of viscous damping devices. They exemplify with pedestrian bridges now equipped with fluid viscous dampers, as the Millennium and the Simone de Beauvoir footbridges.

Chapter 6

The Simone de Beauvoir Footbridge between Bercy Quay and Tolbiac Quay in Paris: Study and measurement of the dynamic behaviour of the structure under pedestrian loads and discussion of corrective modifications

Wasoodev Hoorpah
MIO, Paris, France

Olivier Flamand
CSTB, Nantes, France

Xavier Cespedes
SETEC-TPI, Paris, France

SUMMARY

The Simone de Beauvoir Footbridge in Paris has been from the early stages designed with a complex damping system to avoid both lateral and vertical pedestrian-induced vibrations.

The experimental assessment of damping levels, object of this paper, has been conducted with a large number of walking tests and an extensive instrumentation.

Different unexpected phenomena came to light through these tests: unexpected looseness in footbridge articulations, solid friction on the expansion joints and sliding bearings. These phenomena affected deeply the behaviour of the first and fifth vibration modes.

After corrective actions have been taken, the damping of mode 1 could be markedly increased.

Experimental surveillance of the dynamic characteristics of the bridge during the first few years has been anticipated, that will permit to detect either structural modification or degradation of dissipative devices.

Keywords: footbridge; dynamic; planning; lateral vibration; response; damping; measurements; surveillance.

6.1   INTRODUCTION

The footbridge spans across the river Seine in Paris in between Bercy and Tolbiac quays it is a footbridge with no intermediate supports. Construction on this elegant bridge began in 2006. In this footbridge the principal of a tensioned arc is used to attain an incomparable slenderness with a span of 190 m. The risk of vibrations of this the 37th bridge in Paris were taken into consideration at an early stage by the architect D. Feichtinger and RFR design office who adopted an approach based on studies carried out on the Millennium footbridge in London. The company Eiffel CM, steel contractor, also had some experience of the treatment of vibrations with the Solferino footbridge in Paris. They were supported by the design office SETEC for the modelling of the structure and the design of anti-vibration measures [1], the companies GERB and Taylor for the supply of damping systems and CSTB for the measurement of vibrations. SETRA provided advice on the choice of dynamic testing of the bridge. The site was designed to be a place of passage as well as a place of contemplation of the river Seine and a place were temporary events could be held. The specifications stipulated that the "suspended square" could welcome several hundred people. This potential crowd and its interaction with the dynamics of the structure were the central preoccupation of the design teams, construction teams and the general public.

6.2   PRINCIPLE OF THE SURVEILLANCE MEASUREMENTS

The surveillance of the structure is based on a network of sensors (Figure 6.1, Table 6.1), a data bus and a unit for analysing/saving the data. This system should allow the detection of vibrations of large amplitude, and should be capable of following the dynamic characteristics of the structure and its system of dampers; it should also produce an alert when the limits are passed. The system should function for several years; the greatest care was taken when choosing the sensors and their installation.

Accelerometers were chosen to measure the movements of the aerial parts of the structure. The first mode was calculated to be at a frequency of 0.46 Hz, the sensitivity of the accelerometers has to be very high to take into account the displacements of weak amplitude. The arrangement of the sensors was optimised to limit their number while taking into consideration the deformations by the first 12 modes; these correspond to the range of frequencies likely to be excited by the users of the bridge.

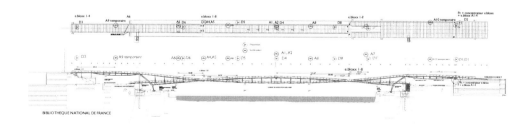

Figure 6.1 Plan of instrumentation of the bridge.

Table 6.1 List of sensors for the surveillance system.

N°	Denomination	Type of sensor	Position on bridge	Direction of measurement	Modes identified	Sensitivity
1	A1	Accelerometer	Central/lateral sup	Transversal	1,5	0.2 g/V
2	A2	Accelerometer	Central/lateral sup	Vertical	**2,5**,9,10	0.2 g/V
3	A3	Accelerometer	Between file 19 et 20/lateral sup	Vertical	2,**4**,9,10,11	0.2 g/V
4	A4	Accelerometer	File 17/lateral	Transversal	1,6	0.2 g/V
5	A5	Accelerometer	File 17/lateral	Vertical	2,4,**6**,8,9,11	0.2 g/V
6	A6	Accelerometer	Between file 14 and 15/lateral sup	Vertical	6,**8**,**10**,11	0.2 g/V
7	A7	Accelerometer	Between file 33 and 34/lateral sup	Vertical	6,**8**,**10**,11	0.2 g/V
8	A8	Accelerometer	File 28/lateralsup	Vertical	18	0.2 g/V
9	A9 Temporary	Accelerometer	File 7/lateral	Vertical	18	0.2 g/V
10	A10 Temporary	Accelerometer	File 41/lateral	Vertical	16	0.2 g/V
12	D3	Displacement	Entrance BNF bank	Transversal	1,5 and dampers check	5 mm/V
13	D2	Displacement	Entrance Bercy bank	Longitudinal	2,5,9,10,16 and dampers check	10 mm/V
14	D1	Displacement	Entrance Bercy bank	Transversal	**1**,**5**,**6** and dampers check	5 mm/V
15	D4	Displacement	TMDcentral/sup	Vertical	9	10 mm/V
16	D5	Displacement	TMD between files 20 et 21/inf	Vertical	11	10 mm/V
17	D6	Displacement	TMD file 15/inf	Vertical	10	10 mm/V
18	D7	Displacement	TMD between files 33 et 34/inf	Vertical	12	Not placed
19	D8	Displacement	TMD between files 30 et 31/sup	Transversal	6	Not placed

Displacement sensors of type LVDT were placed on the different types of dampers, ADA (tuned mass dampers) and viscous dampers, to follow the amplitude of their path. The bandwidth of the data bus (40 Hz) and sensors (0–100 Hz) were chosen according to the range of frequencies concerned with pedestrians walking on the bridge, 0–3 Hz. The bus allows a counter of the number of pedestrians crossing the bridge to be added later on if needed.

This system of surveillance evolved during the study due to the identification of the initial mode of the structure: the number of dynamic dampers used was limited to three; two displacement sensors which were anticipated in the beginning were

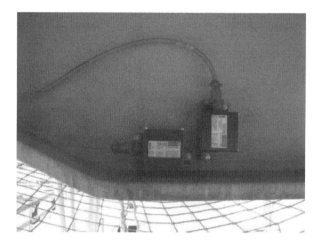

Figure 6.2 Sensors placed on the structure, accelerometers on the left, LDVT sensor on a tuned mass damper on the right.

not installed. Two accelerometers which were meant to be used temporarily and dismounted after the crowd tests were finally left with the permanent sensors.

The measurement of vibrations is permanent; the storage of the data is parametrical.

Moreover two functions were programmed for the data recording system based on a micro-computer linked to the Internet:

– The daily surveillance of the structure is based on the definition of limits.

Instantaneous limits or difference limits which if passed for at least one sensor triggers the recording of a sequence of five minutes thereafter. The measurements for the day are transferred automatically each night to a CSTB server The measurement of the characteristic modes or the damping of the structure on the occasion of punctual tests uses a manual command to record the data.

This may be operated at a distance via the internet or directly from a computer situated on the right bank.

The sensors are positioned in such a way so as to take into account the exposition to climatic factors and to minimise vandalism. The visual impact was also minimised, the cables were placed during the construction phase of the structure (Figure 6.2).

At the extreme right bank of the structure a cupboard houses the computer which controls the data measurements and the internet connection (Figure 6.3).

6.3   SPECIFIC TESTS

Different types of dynamic specific tests were carried out successively on the structure with different aims:

Identification of the different modes of the structure was carried out before the design of the tuned mass dampers; this was done to provide the necessary data for their dimensioning.

Figure 6.3 Acquisition system installed at the extreme right bank.

A second modal identification was carried out after the different systems had been put in place to validate their impact. Other modal identification could be scheduled during the life time of the structure to follow its evolution.

A series of test with a crowd of people were carried out before the inauguration, this was to verify the pedestrian load model and at the same time verify the safe behaviour of the bridge.

Several excitation sequences by an organised group were carried out after the 13th of July 2006, the inauguration date, to verify the response of the structure to an act of vandalism as well as to regulate the system of dampers.

This optimisation period of the dissipative system was necessary as the level of damping was measured inferior than that estimated during the design stage. It lasted until the end of 2006, requiring several structural modifications to the structure after its opening to the public.

6.3.1   Initial identification of modes

The parameters that allow the fundamental natural mode of vibrations to be calculated using a finite element model of the footbridge are the frequency of vibration and its associated modal deformation. Furthermore it is useful to know the modal damping associated with each mode to decide which modes need extra damping by the addition of dissipative devices.

This idea of damping is the dissipation of energy by the structure, which is a non-linear phenomenon as a function of the amplitude of the vibration. It was judged to be pertinent to inject the structure with energy of the same order of that which will be dissipated during its service; this was achieved during tests using an inertial excitation machine (Figure 6.4).

The same machine used to represent the affects of pedestrians on the Solferino footbridge was used. This machine reproduces a sinusoidal force of several hundred

Figure 6.4 Inertial exciter in use for the identification of mode 1.

Table 6.2 Results of the initial identification of modes.

Mode	Calculated frequency (Hz)	Measured frequency (Hz)	Ratio frequency measured/calculated	Critical damping ratio
1	0.458	0.555	1.21	0.56%
2	0.688	0.717	1.04	0.28%
3	0.959	–	–	–
4	1.014	1.042	1.03	0.34%
5	1.122	1.154	1.03	0.28%
6	1.42	1.526	1.07	0.17%
6 bis	1.49	1.526	1.02	0.17%
7	1.56	–	–	–
8	1.601	1.774	1.11	0.18%
9	1.678	1.642	0.98	0.21%
10	2.076	2.139	1.03	0.14%
11	2.116	2.209	1.04	0.31%
12	2.277	2.349	1.03	0.17%

Newtons horizontally or in vertically at a precise frequency. The function of this machine is based on 4 counter-rotating eccentric masses, for the low frequencies a different layout is used to increase the force in the horizontal direction.

Another method allowing the identification of modes consists of measuring the vibrations of the structure under environmental excitation, meaning from uncontrolled excitation produced by the wind, the vibrations of the ground on the bridge supports, from the work in progress and the traffic on the structure. However, the measurement of structural damping is not representative of the structure in service as the level of stresses is normally very weak. In this case this method was used for a first estimate of the modal frequencies to save time when using the inertial exciter machine. After estimating the modal frequencies the inertial exciter machine with the unbalancing mass could quickly sweep through these frequencies find to find more accurate value.

The tests were carried out on the 16, 17 and 18th of May 2006 on the structure in its final phase of construction. The wood decking had not been placed and some point masses due to continuing construction were present on the structure, but their influence on the dynamic behaviour of the structure was judged to be negligible; The risk of excitation by a uncontrolled source was also reduced as the site was not open to the public and meteorological conditions were very calm.

A sequence of measurements was completed with 9 accelerometers placed on the structure (Figure 6.5). These were in addition to the 10 of the surveillance system.

The comparison of the results measured with the modes calculated beforehand shows a good correlation (see Table 6.2). Most of the modes have the deformation anticipated with a modal frequency within a few percent of those predicted by the calculation. Two exceptions however: the modes 8 (vertical 3) and 1 (lateral 1) were identified with frequencies markedly superior than those predicted. Furthermore two modes predicted by the calculations were not identified on the structure, the lateral modes 3 and 7 which concern only the access walkway.

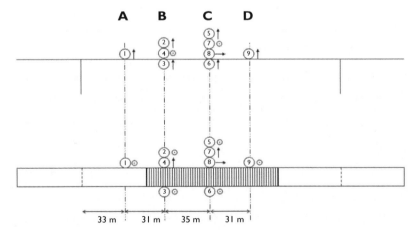

Figure 6.5 Positioning of temporary accelerometers for the measurements of the deformed modes.

Table 6.3 Summary of the modal identification in the presence of dissipative systems activated or not.

Mode	Deformed shape	Frequency (Hz)	Critical damping ratio (%)
1 undamped	OK	0.557	0.58
1 damped	OK	0.562	0.77
2 with TMD	OK	0.706	0.38
5 undamped	OK	1.124	0.53
5 damped	OK	1.134	0.58
6 without TMD	OK	1.384	0.91
6 with TMD	OK	1.485	0.95
8 with TMD	OK	1.730	1.51
9 without TMD	OK	1.626	0.35
9 with TMD	OK	1.613	1.70
10 with TMD		2.115	1.20
11 without TMD	OK	2.170	0.52
11 with TMD		2.180	0.70

6.3.2 Choice of dissipative systems

As a result it was decided that the hydraulic dampers placed at the extremities of the footbridge arrest efficiently the vibrations of the modes 1, 5, 6, 7 and 8. For the modes 9, 10 and 11 it was decided that additional tuned mass dampers would be used.

6.3.3 Identification of modes with dissipative systems

After the dissipative systems, ADA and viscous dampers, had been put into place on the structure, a program of the measurements of the modal dampers was carried out with the same inertial exciter machine that was used to identify the initial modes.

Even though the modal frequency was verified on this occasion, it is modal damping which is the most important. Tests were performed by activating/deactivating successively each damping device put in place, to judge the additional modal damping provided by each on each mode susceptible to be excited by pedestrian users.

From the analysis of Table 6.3 one can conclude that the effect of the viscous dampers placed at the extremities of the structure for the modes 1 and 5 is limited whereas the effect of the dynamic dampers tuned to their target modes is very good. The majority of the damping of mode 8 was found to come from the ADA intended to damp the mode 11.

Finally the vertical mode damping was judged to be very satisfactory. On the other hand the damper for mode 1 was proven to be too weak to minimise the risk of the bridge being put in vibration under the affect of the slow walk of a crowd, called <<Millennium effect>>, this threatened to delay the opening if the bridge to the public.

6.3.4 Tests with a crowd

An important factor of the excitation phenomenon of the footbridge by pedestrians is the synchronisation of pedestrians with the footbridge.

The crowd the most likely to interact strongly with the bridge is not the crowd the most numerous. In fact above a certain number of individuals, the footsteps are uncorrelated and the overall force is perturbed rather than organised.

Table 6.4 List of tests carried out on the morning of the 10th of July 2006.

Code	Characteristics of motion	Aim	Type of walking	Trajectory	Start hour
MaL	slow random walk	Sensitivity to excitation of lateral mode 1 or 2 and torsion 3	Random in group of 20, 40, 60, 80 et 120 persons	Crossing of the bridge with 20 p. then turn back adding 20 each time	10H00
McL	Slow synchronised walk	Sensibilité à l'excitation du premier modelatéral	Synchronised with groups of 120, 80, 60 et 20 persons	Crossing of the bridge with 120 p. then turn back adding 20 each time	10H25
McM	Medium speed synchronised walk	Sensitivity to excitation of torsion/bending mode	Synchronised in groups of 20, 40, 60, 80 and 120 persons	Crossing of the bridge with 20 p. then turn back adding 20 each time	11H00
MaM	Medium speed synchronised walk	Sensitivity to excitation of torsion/bending mode	Random in group of 120 and 40 persons	Crossing of the bridge with 120 then return with 40	11H25
MaR	Fast random walk	Sensitivity to excitation of vertical modes 3 et 4	Random in group of 40 and 120 persons	Crossing of the bridge with 40 then return with 120	11H35
McR	Fast synchronised walk	Sensitivity to excitation of vertical modes 3 et 4	Synchronised in groups of 120 persons	Crossing of the bridge with 120 persons	11H45
Ca	Random running	Sensitivity to excitation of vertical modes 3 to 6	Random with120 and 40 persons	Crossing of the bridge with 120 persons then return with 40 persons separating in two groups	11H50
Cc	Synchronised running	Sensitivity to excitation of vertical modes 3 to 6	Synchronised with 40 then 120 persons	Crossing of the bridge with 40 then return with 120	12H00
MaN	Random normal walk	Sensitivity to excitation of lateral modes 1 or 2 and torsion 1	Natural in group of 120 persons	Crossing of the bridge from left to right bank	12H08

For this reason tests were performed with a crowd of 120 people on the 10th of July 2006. Table 6.4 summarises the various scenaria. During the inauguration of the bridge about 400 people walked along the bridge, without the vibrations measured by the surveillance system exceeding the level attained during the tests.

The aim of these tests was to represent a number of configurations types: the number of people circulating on the bridge was varied from 20 to 120, the path taken followed a number of different forms, the walking rhythm were regulated and covered a range, from the slow walk of a cortege to a group running rapidly.

A predominant parameter for the effect of a group of walkers rests in the cohesion of the group which causes a synchronisation of footsteps. The form of the groups of

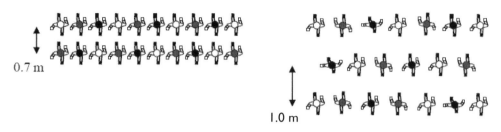

Group of synchronised walkers   Group of random walkers

Figure 6.6 The organisation of groups of walkers has an influence
on the coherence of the forces transmitted.

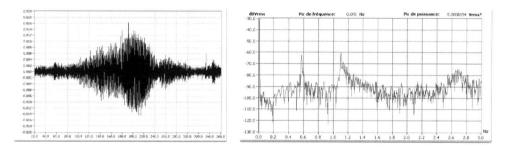

Figure 6.7 Example of a signal from a transversal accelerometer in the middle of the
bridge deck recorded during the passage of 120 people walking slowly
and spectral analysis showing the supremacy of mode 1.

pedestrians participating in the experiment was studied in such a way so as to increase
the coherence in the case of synchronised walking (with the guide of a metronome)
and to reduce this coherence in the case of random walking (see Figure 6.6).

For these tests the displacements of the structure were found using the surveillance system of the bridge and a chain of supplementary accelerometers. These results
were compared to the security criteria and the level of user comfort adopted for the
bridge.

The response of the footbridge to a mobile non-deterministic system of forces
is inevitably complex (see Figure 6.7), but the participation of different vibrational
modes conformed to the predictions of the calculation model. On this occasion the
behaviour of the dissipative devices was also verified in extreme conditions. The
amplitudes read on the tuned mass dampers (ADA) were significant, showing their
usefulness, without passing their capacity.

The maximum amplitudes read were of the order of 30 mm. They correspond to
a excitation of mode 1 by a crowd of 80 or 100 people walking in a random manner (Figures 6.8, 6.9) or by a group limited to 20 people walking in synchronisation.
It was again demonstrated that above a critical number of walkers the amplitude of
excitation decreases. The most important excitations are those caused by a group limited to 20, 40 or 60 people with synchronised movement using a metronome.

Figure 6.8 Random walking of a grouped crowd.

Figure 6.9 Random running of a group of volunteers.

6.3.5   Excitation test with a synchronised group

The tests with a group whose movements are synchronised by a metronome produced high excitation amplitudes. Also for the excitation of mode 1 a group of 60 people succeeded in causing on the10th of July a amplitude of sway equal to 60 mm in the middle of the bridge deck, double the effect of a crowd twice as numerous.

The easiness of the test meant that it could be repeated five times in the months that followed the opening of the bridge, in order to test the efficiency of the structural modifications with the aim of increasing the damping of mode 1.

This type of test was repeated on the 18th of July, the 22nd of August, the 23rd of August, the 9th of November and the 28th of March 2007. The method was refined due to the focus of the tests on the structural damping of the 1st mode. The excitation of this mode is easily achieved by a group of about 30 people in the middle of the central span of the bridge, on the lower part. By making the volunteers sway from one foot to the other at a frequency double that of the 1st mode, being 1.12 Hz, one can excite this mode in around ten seconds. The group was made to sit down straight away so that the standing up position did not add to the damping <<biological>> of people to the structural damping of the footbridge. From signal recorded by the sensors

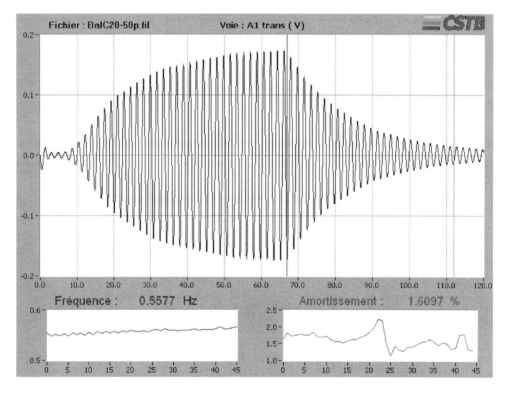

Figure 6.10 Acceleration in the mid span for an excitation by a group of 60 people.

of the surveillance system or the additional sensors we can deduce after filtering the bandwidth the damping of the structure for the different amplitudes of vibration, until the suppression of the oscillatory movement (Figure 6.10).

Different phenomena came to light by these tests: play (looseness) unexpected in the footbridge articulation on the extreme right bank, solid friction in the expansion joint at the extreme left bank (Figure 6.11), and the right bank as well as for sliding bearings of the access footbridge were found from different amplitudes of vibration on different points of the structure, and then corrected. The damping of mode 1 was markedly increased.

The dependence of the damper for mode 1 with the amplitude of oscillations was also noticed (Figure 6.12).

The last tests showed a level of damping satisfactory for the average amplitudes which increase if the amplitude of the vibrations should amplify. The stability of the structure is sufficient to reduce the risk of an interaction between the structure and the public as in the case of the Millennium Bridge.

Surveillance of the dynamic characteristics of the footbridge during its first few years of use has been anticipated. This will permit detection of a shift of modal

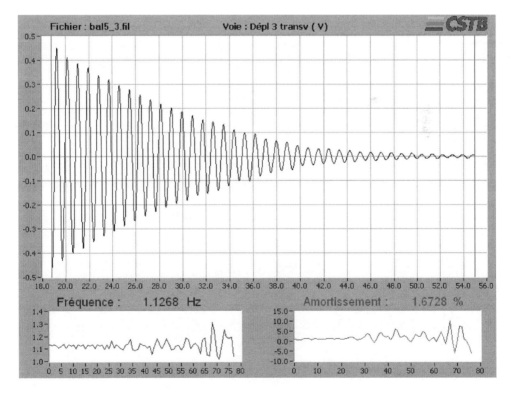

Figure 6.11 Example of displacements measured on the 23rd of August 2006 at the extremity of the left bank of the structure showing solid friction.

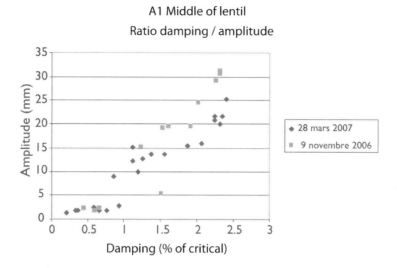

Figure 6.12 Dependence of damping of mode I with amplitude of vibration.

frequency of or a change in the damping due to, structural modifications, degradation of the structure or wear and tear of dissipative devices.

This technique has been used with success to understand the reasons of initial weak damping of the 1st transversal mode and to correct the structure to obtain satisfactory damping, with out a doubt this will be useful in assuring the security of the structure in the long term.

REFERENCES

[1] Cespedes X, Flamand O, Manon S., "**Passerelle Simone de Beauvoir - l'ouvrage sous charges piétonnes**", Ouvrages Métalliques, 01/01/08, n° 5, pp. 52–63, 2008.

Chapter 7

The influence of dynamics in footbridge design: North American practice

Theodore Zoli
HNTB, New York, USA

The excessive sway motion of the Millennium Bridge in London has focused a great deal of attention on serviceability issues associated with Pedestrian-Induced Vibration (PIV). A recent CEB-FIP publication *Bulletin 32 Guidelines to the Design of Foot-bridges (November, 2005)* devotes an entire chapter to dynamics but includes little or no discussion of any other aspect of footbridge dynamics (wind, seismic, member loss). While PIV represents a critical aspect of design and dominate current footbridge research, a number of recent major footbridges in the US have been very much influenced by other dynamic design considerations.

This paper explores aspects of non-pedestrian induced dynamics on footbridge design, and the interplay between i) competing design requirements with a focus on footbridge dynamics, ii) superstructure typology, iii) constructability and iv) construction cost and project viability.

For this paper, three recent US bridge projects, the Missouri River Pedestrian Bridge, the Mary Avenue Pedestrian Bridge, and the Happy Hollow Pedestrian Bridge are of particular interest and will be explored in some depth. All three footbridges share a common heritage; each bridge as initially conceptualized and designed has been bid in a competitive process, each project has been significantly over-budget (more than double the project budget), and had been cancelled. In each case, the bridges have been redesigned with a clear focus in the redesign effort on cost savings based upon enhanced dynamic performance.

7.1 OVERVIEW

As can be seen from Figure 7.1[1], footbridges of significant span length, regardless of bridge type and superstructure material, are potentially susceptible to vibration serviceability

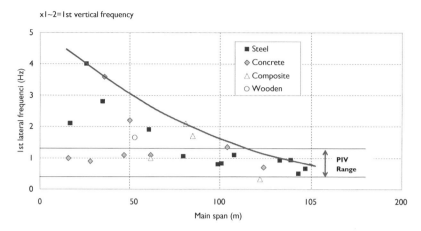

Figure 7.1 Fundamental lateral frequencies of various bridges by type & span length.

issues associated with Pedestrian Induced Vibrations (PIV). It is therefore not surprising that PIV is at least a major design consideration in conceptualizing footbridges that exceed 80 m in main span length. Three design strategies employed for footbridges susceptible to PIV include [2]:

• Frequency Tuning – Avoid the critical PIV frequencies with the fundamental modes of vibration
• Perform a Detailed Vibration Assessment
• Implement Measures to Reduce Vibration Response (added damping, or access restriction).

Each design strategy against pedestrian induced vibration has intrinsic disadvantages, particularly from the perspective of overall bridge performance against the wide array of governing bridge design loads. Design strategies that significantly reduce the potential for PIV serviceability issues may have such negative impacts to bridge design complexity, constructability, construction cost, and even safety as to push a footbridge project past the point of financial viability given typical budgetary restrictions of many North American bridge owners.

It is the intent of this paper to use three recent North American bridge projects to highlight circumstances where footbridge dynamics beyond PIV have been a major design consideration in order to highlight the interaction and in some cases the conflict between PIV and other key dynamic design considerations.

The paper is organized in the following manner. To begin, each PIV resistant design strategy is presented with a brief discussion of where there is both interaction and conflict with other aspects of footbridge design. Next, three project case studies are presented wherein design strategies that highlight aspects of footbridge dynamics and how they impacted design development are considered.

Finally some conclusions are drawn regarding footbridge dynamics as they impact footbridge design.

7.2   FREQUENCY TUNING

Frequency tuning to avoid potential resonance with PIV is clearly an appropriate and feasible strategy for footbridges where span length and bridge form are not dictated by site constraints. However, for cable supported footbridges with main spans that exceed 75 m, such a strategy is rarely viable given typical walkway geometries, span configurations, and bridge typologies (arch, cable stayed, suspension, stress ribbon). Tuning strategies which keep the fundamental vertical modes outside of the 1.6–2.4 Hz range and lateral modes outside the 0.8–1.2 Hz can present significant problems for footbridges in regions of high seismicity as is outlined below.

High seismicity in the west coast of North America California presents a distinct challenge to frequency tuning as a design strategy. Figure 7.2 depicts smoothed acceleration input response spectra for the Mary Avenue footbridge project site in Northern California. The design spectral curve (solid line) represents an envelope of the 4 identified contributing spectra associated with faults in sufficient proximity to the project site. As is readily evident, frequency tuning to avoid PIV has the consequence in shifting bridge response toward a period range that maximizes seismic response; thereby maximizing force demands and potential vulnerability to seismically induced damage. Shifting the fundamental horizontal modal behavior above 1.2 Hz (0.83 seconds) and vertical modal behavior above 2.4 Hz (0.42 seconds) may very well be inappropriate for bridges in areas of high seismicity. This has been a particularly important consideration in the conceptual design of

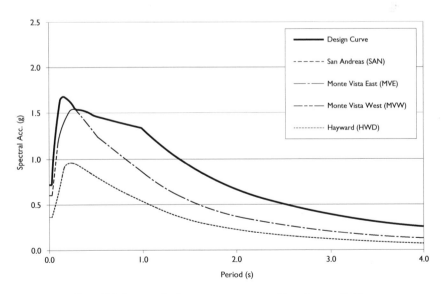

Figure 7.2 Input seismic response spectra, Mary Avenue Bridge.

both the Happy Hollow and Mary Avenue Pedestrian Bridges as will be discussed herein.

7.3   DETAILED VIBRATION RESPONSE ASSESSMENT

An outcome of a wealth of recent research in PIV has at least design guidance based upon conducting a detailed vibration response assessment for serviceability design. As put succinctly by Zivanovic, et al [2] in their comprehensive literature survey on PIV of footbridges, "this is easier said than done", as there remain many issues regarding such assessment methods.

Footbridge susceptibility to PIV is strongly influenced by the following

- Modal vibration characteristics of the bridge
- Modal damping
- Applied load models and excitation type (Walking, Running, Jumping)
- Pedestrian densities
- Pedestrian distribution along the footbridge
- Pedestrian Footbridge Interaction (degree of synchronization & lock-in)
- Comfort criteria (stationary versus non-stationary pedestrians

Over the past few decades there have been significant advances in the modeling of bridges, particularly the use of the Finite Element Method (FEM) to predict bridge modal characteristics. However, in many cases FEM modeling may not be sufficiently accurate for the purposes of assessing serviceability concerns related to PIV. A recent paper compares FE Model and modal testing for a relatively simple steel box girder footbridge in Montenegro, whereby the first asymmetric vertical mode was underestimated by 37% [3] (pre-test FEM estimate 2.11 Hz, Measured 3.36 Hz). The contribution of handrails and other non-primary structural elements also may have a significant impact on bridge dynamic characteristics.

PIV susceptibility is highly dependent upon modal damping, since it is a near resonant phenomena. Unfortunately, modal damping can only be determined with sufficient accuracy following completion of the bridge (not much help to the

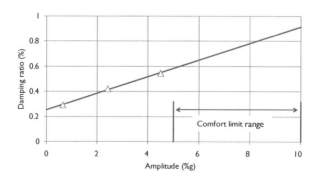

Figure 7.3 Damping & amplitude dependence.

designer). Damping is known to be dependent upon many parameters including material type, structure configuration, articulation, boundary conditions, and even the presence and distribution of pedestrians on the bridge. Forced vibration tests of the I235 arch footbridge in Iowa showed a very clear relationship between damping and amplitude as depicted in Figure 7.3 for the vertical mode with a measured frequency of 2.34 Hz [4]. While recommendations for a range of expected damping level exist, such as has been provided in the CEB-FIP *Guidelines in the Design of Footbridges*, or by Bachmann et al [5] based upon data collected from over 40 footbridges, there remain many uncertainties associated with assigning an associated damping, particularly the dependence upon amplitude as it relates to PIV.

There has been a significant body of work in determining pedestrian induced loads, particularly from an individual pedestrian, together with the development of functional time-domain deterministic and probabilistic force models, for walking, running, and jumping in the vertical, lateral and longitudinal directions. There is less general agreement in terms of modeling the behavior of pedestrians in small groups, as compared to a steady stream of pedestrians (together with a pre-determined density and spatial distribution associated with such a stream). Complexity in the dynamic interaction between footbridge and stationary/non-stationary pedestrians is not well-understood, nor is the degree of synchronization that may be anticipated in both vertical and lateral directions.

Finally, there is little agreement to what may be used as a definitive comfort criterion for pedestrian induced vibration, particularly for vertical vibration limits. A recent paper by Kasperski [6] compares acceleration limits from 8 different sources which vary be nearly an order of magnitude. In the CEB-FIP *Guidelines for the Design of Footbridges*, the following guidance is offered on vertical acceleration criteria,

> "the allowable values for vertical criteria vary greatly between different international standards and in the literature. The authors emphasize that these values in the codes, even though they vary, are conservative. There are footbridges that exhibit vertical accelerations above 1.0 m/s² when excited by two pedestrians or vertical accelerations greater than 1.5 m/s² when passed by two joggers that have never caused complaints."

It becomes hard to justify the need for PIV suppression measures for a footbridge during the design phase unless there are very clear indications of PIV susceptibility to a high degree. It appears prudent to conduct a vulnerability assessment during the design phase, and to adopt design strategies that reduce the potential for PIV serviceability issues but to limit a detailed assessment until after substantial completion of construction.

In summary, the performance of a detailed vibration assessment in the design phase is only of limited use, given the inherent unknowns associated with the dynamic behavior of the bridge once constructed (vibration characteristics, damping), the uncertainty in aspects of pedestrian loading (particularly degree of synchronization, dynamic interaction with the structure, pedestrian density, etc) and finally definitive criteria to assess the potential for PIV serviceability issues. Such a detailed vibration assessment seems more prudent following construction, where the unknowns are substantially reduced and liveliness can be measured directly.

This suggests that an over-emphasis in the design phase of PIV, or more properly an underestimate of the importance of other dynamic design considerations may lead to imprudent design.

7.4   IMPLEMENT MEASURES TO REDUCE VIBRATION RESPONSE

There are a number of highly effective strategies for reducing PIV serviceability by enhancing structural damping, primarily through the use of passive dampers, with the use of tuned mass dampers (TMD's) and tuned liquid mass dampers (TMLD's) preferred particularly when only a single mode need be damped. Combinations of viscous, visco-elastic, and friction dampers may be necessary if multi-modal damping is necessary.

Pendulum type TMD's are the preferred strategy for PIV mitigation due to their proven effectiveness, adaptability, relatively low cost, and ease of implementation. It is noted that vibration tests are a vital aspect of TMD commissioning given the need for precise tuning to the mode to be damped. TMD's must also be accessible for adjustment in most circumstances.

One key concern regarding TMD's, is that they add to the generalized mass for non PIV dynamic events and therefore enhance bridge vulnerability to damage during an extreme event. In addition, it would appear likely that TMD's must include buffers in order to avoid damage during a wind or seismic event wherein large amplitude oscillations may be anticipated.

7.5   CASE STUDY 1 – MISSOURI RIVER PEDESTRIAN BRIDGE, OMAHA, NE

A crossing of the Missouri River presents a dramatic opportunity for a pedestrian bridge, particularly given the span, and the proximity of the bridge to downtown Omaha, Nebraska, a major urban environment in the central United States. An original scheme developed by Bahr Vemeer Haeker Architects (BVH) and Figg Bridge Engineers (FBE) adopted canted towers supporting an S-curved precast segmental 3 m deep concrete box superstructure with a single plane of stays along the inside of the curve. It is presumed that the use of the segmental concrete box superstructure was in part to limit the potential for PIV. However given the large structure depth and overall mass, this superstructure type resulted in the need for unusually large stay cables, pylons and foundations for a footbridge of this span and configuration.

The BVH/FBE project was bid and the lowest responsive bidder ($45 million) was over twice the original budget. The project was re-advertised as a design build with a specified maximum cost of $22 million. The primary basis for selecting the winning scheme was aesthetics. This project delivery strategy of best design to a fixed price is an unusual but challenging format for design/build teams. Three competitive entries were submitted and the team of APAC/HNTB was selected for this crossing.

Given the severe budgetary restrictions for the project, our initial type studies concentrated on tangent alignments for the main span across the Missouri River.

Figure 7.4 BVH/FBE design.

However, the alignment of the bridge, together with the environmental and connectivity constraints for the Omaha and Council Bluffs river edges, pointed toward retaining the S curvature of the BVH/FBE concept. We developed an S shaped bridge on nearly the same alignment and with the same main span configuration as the original design. It was necessary to design an identically configured bridge to the BVH/FBE design for roughly 40% of the as-bid cost. To accomplish this, we set the following goals with a primary focus on cost-effectiveness:

1   Develop the most efficient superstructure possible, minimizing materials (weight) and structure depth (wind).
2   Minimize design complexity and three dimensional behavior with a structural system optimized for in-plan curvature. The cable stayed form was ideal for this span, however two planes of cables were necessary in order to enhance efficiency and reduce design/construction complexity.
3   Minimize geometric complexity in fabrication and erection by ensuring symmetry and repetitive field sections.
4   Minimize temporary falsework/temporary support requirements for erection. Minimize foundation costs to the extent possible.
5   Simplify the dynamic behavior and provide a structural system with proven past performance in resisting wind-induced vibration. Minimize the potential for PIV to the extent possible.

It is noted that wind was the governing load for the crossing and the potential for wind instability represented the most significant risk since time would not permit a wind tunnel evaluation of the cross-section during the bid period and suppression measures to reduce flutter/vortex shedding excitation or significant impacts to pylon and foundation design would result in a project that could not meet budgetary limits.

7.6   SUPERSTRUCTURE CROSS-SECTION & CABLE ARRANGEMENT

While many possibilities exist for curved superstructure alternatives, geometry control, high fabrication costs, and difficult erection are typically anticipated. In order to develop a curved superstructure alternative for this crossing, we fundamentally had to develop a structural system that was easy to erect and to control geometry, cost-effective to fabricate, stable under wind and could resist the lateral forces induced by curvature.

To enhance constructability and to reduce fabrication & erection costs, we decided that the superstructure field sections would be fabricated from rolled universal beams and be chorded (piecewise straight), with angle breaks to form the desired curvature only occurring at splice locations. Given circular horizontal geometry, all field sections were designed to have identical geometry with a constant angle break at each connection. Further, connections were optimized such that girder field splices also incorporated the cable and lateral bracing connections.

Lateral bracing consisted of bolted K type configuration throughout the length of the superstructure. This provides the optimal bracing arrangement for the edge girders which are non-composite, and significantly enhances lateral stiffness.

7.7   AEROELASTIC BEHAVIOR – SUPERSTRUCTURE

Optimizing bridge dynamics under wind excitation (aero-elastic behavior) is often the most challenging constraint in long span bridge design, since poor dynamic behavior (flutter instability and/or vortex shedding excitation susceptibility) requires expensive mitigation measures or significant changes to the cross-section. We utilized parametric studies of edge girder cross sections by Lin et al, [7] to determine an optimal girder depth to height ratio to ensure a stable Flutter Derivative (A_2^*) We selected a structure width of 7.5 m, comprised of universal rolled beam edge girder sections with a depth of approximately 0.58 m (W21's), resulting in a ratio of B/H approaching 13.

To enhance resistance to vortex shedding excitation, we also incorporated a gap between the deck slab and the edge girder to allow venting which disrupts the pressure differential associated with vortex formation in bluff body aerodynamics. This has also been shown to be an effective means of enhancing a cross-section's resistance to flutter [8].

Section model tests were completed during final design and are depicted in figures 7.5 & 7.6. It is noted that this combined strategy of a large B/H ratio combined with a gap between girder and deck panel were extremely effective at producing a cross section that is not susceptible to flutter instability or vortex shedding excitation.

A key aspect in determining field section length and cable spacing was an evaluation of structural behavior under cable loss scenarios. Based upon this evaluation, field sections (and cable spacing) of 7 m in length were chosen as optimal. Girder slenderness (depth of the superstructure with respect to the main span) is 154/0.58 = 266, well below the maximum slenderness ratios used on pedestrian and vehicular cable stayed bridges which exceed 300.

The gap between girder and slab required that the deck is not composite with the edge girders. While this is unusual for cable stayed bridges, it has a number of intrinsic

Figure 7.5 Missouri River Pedestrian Bridge.

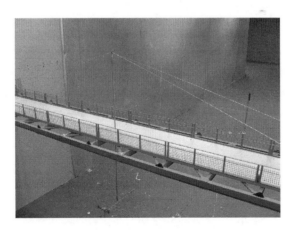

Figure 7.6 Section model test.

advantages for footbridges beyond enhanced wind performance. First and foremost, a non-composite deck allows for full deck replacement. This will be the first U. S. cable stayed bridge ever where the deck may readily be replaced, a fundamental advantage in terms of maintainability. The precast panels were sized to be transportable over the road (7 m long by 5 m wide) with special permits.

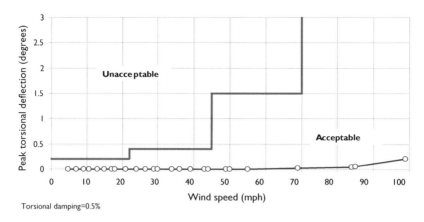

Figure 7.7 Section Model Behavior for Vortex Shedding & Flutter.

This unique arrangement of a precast non-composite concrete deck also provided a potential opportunity to enhance resistance to PIV by incorporating the elastomeric bearings as a constrained layer damper between the walkway surface and the supporting structural steel. In addition, compression seal type expansion joints will be used between deck panels to potentially further enhance superstructure system damping. The degree to which this arrangement results in enhanced damping will be known only after the bridge is completed, but this superstructure arrangement, developed primarily to provide optimal aeroelastic characteristics, suggests an alternative design strategy to reduce PIV susceptibility.

7.8   PYLONS AND PYLON FOUNDATIONS

The single mast pylon cross section and single large diameter drilled shaft foundations were selected for reasons of economy, constructability and performance under vessel impact. A single 4 m diameter drilled shaft with a 3.5 m diameter rock socket was selected as the most efficient structural system to resist vessel impact design loads minimize scour and support dead and live loads associated with the cable supported main span unit. A single drilled shaft eliminated the need for cofferdam and tremie seal construction associated with a pile cap and the 24 1 m drilled shaft foundation system in the BVH/FBE design, significantly reducing construction risk, expense, and potential schedule delays associated with adverse river conditions.

Another fundamental premise in our design is that the pylons be vertical and positioned such that they are coincident with the center of mass of the superstructure in plan. Thereby transverse moment in the pylons under dead load is eliminated. The curvature of the superstructure and the position of the tower were chosen specifically to achieve this concentric geometry. Past experience with inclined concrete pylons suggested that the large creep & shrinkage induced displacements associated with inclined towers were inappropriate for this project.

Figure 7.8 Pylon anchor head concept.

The triangular cross sectional shape of the pylon was selected for the following reasons:

- Constructability – a triangular shape allows the opportunity to taper in two directions without complex formwork.
- Structural Efficiency – the triangular shape maximizes sectional strength while minimizing cross sectional area.
- Wind efficiency – the triangular shape with rounded corners is geometrically one of the most efficient shapes.
- Cross Section CG – The CG of the cross section is such that (with the flat side oriented adjacent to the walkway) that the CG of the tower can be as close as possible to the cross-section. This minimizes curvature and the unbalanced condition during erection.

Pylon reinforcement is organized in such a way as to maximize prefabrication. Each corner node is a pre-tied cage that is linked together with overlapping circular ties. This arrangement minimizes the need for difficult to place and in-effective lateral ties as well as standardizing the reinforcement to the degree possible. As the section tapers, the diameter of the circular ties as well as the number of vertical bars within the circular ties are the only reinforcement that changes.

The complex cable geometry associated with the curved superstructure is alleviated with the three tube anchor head. Fin plates are welded to the side of the pipes at precisely the correct angle (in plan). All fin plates are standardized, such that two pieces of information fully describe the geometry necessary for fabrication (elevation and angle in plan). The 10 cables per side on the inside of the curved superstructure attach directly to the single pipe on the pointed side of the triangle in plan, whereas the cables from the outside of the curved superstructure attach to the two pipe sections oriented along the flat side of the triangle.

An unanticipated outcome of the tapered triangular pylon shape was potential susceptibility to galloping above a certain elevation prior to the installation of stay cables. Galloping is associated with a negative slope to the lift coefficient, and the tower was found to be susceptible for wind speeds that exceeded 60 mph over limited angle of attack. Given the potential for instability a tripod pendulum tuned mass damper was designed and deployed during the period of time that the tower was vulnerable to galloping instability. To save on TMD costs, commercial off the shelf car shocks manufactured by KONI were utilized for damping the TMD mass.

7.9   PEDESTRIAN INDUCED VIBRATION

As mentioned above, wind dynamics were the fundamental concern during conceptual design. Given the dynamic characteristics of 154 m main span cable stayed bridge with a relatively light superstructure, it was clearly not possible to avoid the potential for PIV. A vibration assessment was conducted to determine susceptibility to PIV based upon FEM computed dynamic characteristics. A detailed presentation of the analysis method is reported elsewhere [9]; the results of this assessment identified the potential for vertical acceleration beyond comfort criteria in higher modes, since fundamental vertical modal frequencies were outside the range of vertical PIV. The vibration assessment technique identified the minimum number of pedestrians placed for maximum effect to produce accelerations above a threshold as well as the cumulative effect of a steady stream of pedestrians consistent with a sporting event or concert at a nearby venue. It is interesting to note that the modes identified as susceptible are higher modes than have been reported in the literature for other lively bridges. The degree to which pedestrian bridge interaction, the potential for increased damping in higher modes, the spatial distribution of pedestrians, etc impacts susceptibility to PIV is unknown. In addition, the degree to which the elastomeric bearings and compression seals serve to enhance modal damping and reduce the potential for PIV will be

Figure 7.9 Deployment of pendulum TMD to suppress pylon galloping instability.

of interest. A testing program for the Missouri River Bridge scheduled in the next few months as the bridge nears completion will hopefully shed light on these aspects of PIV.

7.10   CASE STUDY 2 – MARY AVENUE PEDESTRIAN BRIDGE, CUPERTINO, CA

The Mary Avenue pedestrian bridge is to be California's first cable stayed bridge, with a span of 100 m main span crossing I-280 in the city of Cupertino. The site constraints dictated short back spans and a cast-in-place on falsework post-tensioned concrete superstructure as well as a cast-in-place concrete A-shaped pylons. Cast-in-place construction is consistent with CALTRANS practice for vehicular bridges, and was therefore a logical choice by the original design team and consistent with design strategies that would act to reduce bridge susceptibility to PIV. Unfortunately, the lowest responsive bidder at $12.6 million, was more than twice the project budget.

Here again the project was cancelled and we were brought in to identify potential strategies to reduce costs while maintaining the overall configuration and look of the bridge. We identified design strategies to simplify/speed up construction, as well as

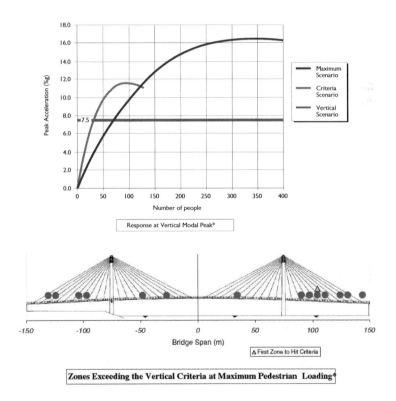

Figure 7.10 Vertical pedestrian vibration assessment – mode 25–2.16 Hz.

to optimize seismic performance while maintaining overall geometry similar to the original design concept. Fundamental to this approach was the use of structural steel pylons and a steel composite superstructure of a unique configuration to enhance seismic response. The design strategy was to achieve a dramatic mass reduction in superstructure and pylons and an overall increase in bridge flexibility to limit seismic force demands. Additionally, we sought to simplify falsework requirements over active interstate traffic, since this aspect of the project introduced a significant amount of risk, construction time and lost costs in falsework erection and subsequent removal. The redesigned structure outlined below was rebid with the low bid at $7.3 million, well within the owner's budget and the project is currently under construction with anticipated completion late this year.

7.11   SUPERSTRUCTURE REDESIGN

The redesigned superstructure utilizes universal steel sections (edge girders and floorbeams) with precast concrete deck panels resulting in a significant reduction in weight in order to obtain synergistic benefits for other proposed cost-savings measures for the pylons, stay cables and foundations. Superstructure mass was reduced by more than 50% leading to major reductions in seismic force demands for pylon, substructure and foundation elements. To ensure adequate aerodynamic stability and based upon the successful experience with the Missouri River Pedestrian Bridge, a gap was incorporated between edge girder and deck slab. The use of a steel superstructure eliminated the need for falsework and significantly reduces the requirements for temporary bents. The superstructure may be prefabricated in 25 m segments and shipped to the site pre-assembled and ready to erect,

Figure 7.11 Mary Avenue Pedestrian Bridge Crossing.

reducing impacts to traffic, enhancing constructability and reducing construction schedule.

High seismic demands precluded the use of the non-composite deck supported on a K braced superstructure as was used on the Missouri River Project. Instead, a strategy to enhance seismic performance be eliminating the K bracing and making the slabs composite with the floorbeams was investigated. The resulting superstructure behaves like an eccentrically braced frame (EBF) a structural system that has enhanced seismic performance typically used in vertical construction. The floorbeams and edge girders are subjected to weak axis bending under lateral seismic loads and the behavior is akin to a vierendeel truss with very short beam elements. This design strategy enhances lateral flexibility and results in a structure with a large degree of inherent ductility in that both beams and edge girders can form plastic hinges.

7.12   REDESIGNED PYLONS

Given the large reduction in seismic demands, the redesigned pylons utilize fabricated steel for the entire height. Use of an I shaped cross section for both tower legs and cross-struts simplify fabrication and cable connections. The pylon is fabricated in halves and bolted at the strut below deck and the tower top to simplify transport and erection. All field connections are bolted; all shop fabricated connections welded. Grade 350 MPa HPS steel was used throughout, to enhance ductility and minimize fabrication costs.

Advantages of this pylon concept included:

- Shop fabricated steel towers
- Simplifies tower construction and cable connections

Figure 7.12 Bridge elevation.

- Reduces construction schedule, erection time and risk
- Grade 350 MPa HPS steel enhances ductility and reduces mass
- Fabricated in halves to simplify transport and erection.

7.13 PEDESTRIAN INDUCED VIBRATION

The revised design is arguably more susceptible to PIV, given the reduction in super-structure mass and increased lateral and vertical flexibility. However, the budgetary constraints dictated the need for dramatic savings and enhanced seismic performance was a specific goal of the redesign.

In order to enhance both seismic and pedestrian induced vibration behavior of the structure, a semi-integral connection of the superstructure to the abutment was developed. This resulted in only a moderate increase in stiffness under seismic load-ing, given the flexibility associated with the abutment foundations (twin 0.8 m drilled shafts) however it results in a significant increase in lateral stiffness under small ampli-tude, shifting the fundamental transverse period outside of the range susceptible to PIV as well as enhancing radiation damping under both dynamic loading scenarios.

Similar PIV sensitivity tests are planned for this bridge as it nears completion, in order to assess liveliness under anticipated pedestrian traffic. Given the unusual super-structure typology and semi-integral abutment configuration, it will be interesting to see how the structure behaves and the overall effectiveness of this structural configuration.

7.14 CASE STUDY 3 – HAPPY HOLLOW PEDESTRIAN BRIDGE

The Happy Hollow Pedestrian Bridge project is in San Jose, within a few miles of the Mary Avenue bridge and therefore subjected to a similar seismic environment

Figure 7.13 Shop fabrication of pylon in half.

and design considerations. The project was originally conceived as a single tower asymmetric cable stayed bridge with a bent alignment and asymmetric spans of 70 m and 90 m. Given the asymmetry and bent configuration, a 55 m tall three legged steel pipe truss tower required two sets of back stays to balance longitudinal forces. The superstructure was comprised of structural steel edge girder and floorbeam system with a wood deck. Again the low bidder for the project submitted a bid of $7.3 million, nearly double the available budget for the project.

To our knowledge, PIV was not a design consideration in the development of this alternative, and given the extreme lightweight and flexibility of the superstructure system, we would anticipate the potential for significant PIV serviceability issues, particularly since the footbridge serves as the main entrance to the Happy Hollow Park and Zoo, and will likely be subjected to significant foot-traffic during special events and on holidays.

7.15 CONCEPTUAL DESIGN STRATEGY

To optimize the potential for the project to be responsive to tight budgetary constraints, we decided to change bridge type entirely and investigate the use of twin network tied arches on a similar alignment with identical span lengths of 85 m. We felt that the network tied arch was the ideal form for this span length is the network tied arch, in that it is the most efficient structural form in terms of materials for this span length. This bridge form was originally developed by a Norwegian engineer named Per Tveit and involves the use of a large number of crossing suspenders instead of the more conventional tied arch that uses widely spaced vertical suspenders. Tveit's Bolstadstraumen bridge is depicted in Figure 7.14.

With the inclined suspenders forming a network of cables, the behavior of the tied arch approaches that of a truss, nearly eliminating bending moments in the arch and tie girder. In this manner, section depths are reduced to the minimum, since the arch and tie girder are acting more like chords of a truss.

An extremely efficient and lightweight bridge superstructure is ideal from a seismic perspective as the overall mass of the system is limited, particularly the superstructure mass which drive seismic demands on the foundations. In addition the use of independent simple spans is ideal for the use of seismic isolation bearings to reduce seismic load demands on the superstructure and substructure.

From the perspective of pedestrian bridge vibrations, the network cable configuration enhances the stiffness of the structural system in vertical, lateral, and torsional dynamic response, and will provide enhanced resistance to such excitation. This is a similar strategy to that proposed for the Mary Avenue Pedestrian Bridge project; to develop a structural system that is adequately stiff for pedestrian induced vibrations but sufficiently flexible such that seismic force demands are not excessive which drives overall bridge cost.

It is noted that a key aspect of network tied arch bridge design given the slenderness of the arch rib and the importance of the suspenders to stabilize the structure is to evaluate system safety under dynamic cable loss, and to consider the effects of multiple cable loss associated with some unforeseen event. Cable loss dynamic analyses have been conducted to establish a preferred suspender spacing of approximately 3 m.

Figure 7.14 Tveit's Bolstadstraumen Bridge – completed 1963.

Figure 7.15 Happy Hollow Pedestrian Bridge proposal.

7.16   SUPERSTRUCTURE DESIGN

From the perspective of efficiency, given that the behavior of this bridge form is akin to a truss and not a conventional arch, it is possible to use a conventional rolled section oriented in weak axis bending for the arch rib (we often use such sections as chords in trusses). It is also unnecessary to use a parabolic shape for the arch, instead a circular shape may be used which allows the arch to be a constant radius, with every section identical. It is also feasible to use a constant section, i.e. a single cross-section for the entire arch. Similar opportunities for optimization exist for the tie girder. In fact, it was decided to utilize a single rolled section for the tie girder and lateral bracing as well, a WT section (cut from a single wide flange section). Thereby, for the entire bridge only two rolled sections are utilized, a W14 × 90 for the arch ribs and a W18 × 76 (cut in half) for the tie girders and cross bracing.

Instead of the use of ironwood decking, a reinforced concrete supported on stay-in-place galvanized corrugated metal deck forms is proposed. This deck system results in substantial cost reduction both in terms of first cost and maintenance cost, and has the added advantage that the composite deck becomes an integral part of the overall structural system. The corrugated metal deck is conservatively designed as a sacrificial element, the deck has sufficient reinforcement for all dead, live, and vehicular loads. The use of stay-in-place rather than removable deck forms is the most cost-effective bridge deck construction strategy.

7.17 PROJECT STATUS

Preliminary design has been completed and it appears that significant cost savings can be achieved with the proposed redesign. Final design is anticipated to begin in the next few months, where it will be necessary to confirm that the proposed design meets or exceeds seismic design requirements and has adequate resistance to Pedestrian Induced Vibration.

7.18 SUMMARY

While pedestrian induced vibration remains a dominant consideration in the design of long span footbridges, wind and seismic behavior often drive design, particularly in high seismic regions in western North America. Design guidelines for footbridges appear to underestimate the potential for seismic and wind design to be major design considerations and to identify the potential conflicts associated with PIV resistant as compared to wind and seismic resistant design strategies.

Case studies of recent pedestrian bridges in the United States have outlined new design strategies for improved wind and seismic performance, while attempting to minimize susceptibility to PIV. The degree to which these design strategies are effective strategies for reducing susceptibility to PIV will be determined in the near future as these projects near construction completion, allowing for a proper assessment of their liveliness.

REFERENCES

[1] Dallard, P., Fitzpatrick, A.J., Fling, A., Low, A., Ridsdill, R.M. and Wilford, M., The London Millennium Footbridge, *The Structural Engineer*, Vol. 79 (22), pp. 17–35, November 20, 2001.
[2] Zivanovic, S. et al. *Vibration Serviceability of Footbridges under Human-Induced Excitation: a Literature Review*, Journal of Sound and Vibration, Vol. 279, No. 1–2, pp. 1–74 (2005).
[3] Zivanovic, et al., *Modal Testing and FE Model Tuning of a Lively Footbridge Structure*, Engineering Structures, Vol. 28 (2006).

[4] Byers, D.D., Stoyanoff, S. and Boschert, J.P. Dynamic Testing of the I-235 Pedestrian Bridge for Human Induced Vibration, *IBC Pittsburgh*, 34, 2004.

[5] Bachmann, H. et al., *Vibrations Induced by People*, Vibration Problems in Structures: Practical Guidelines, Birkhauser Verlag, Basel. (1995).

[6] Kasperski, M. *Vibration Serviceability for Pedestrian Bridges*, *ICE Structures & Buildings*, Vol. 159, Issue SB5, October 2006.

[7] Lin, Yuh-Yi et al., *Effects of Deck Shape and Oncoming Turbulence on Bridge Aerodynamics*, Tamkang Journal of Science and Engineering, Vol. 8, No. 1, pp. 43–56 (2005).

[8] Ehsan, Jones, and Scanlan, *Effects of Sidewalk Vents on Bridge Response to Wind*, J. Struct. Engrg., Volume 119, Issue 2, pp. 484–504 (February 1993).

[9] Stoyanoff, et al., *Pedestrian Induced Vibration on Footbridges: advanced response analysis*, Bridge Structures, Assessment, Design and Construction, Volume 3, Taylor and Francis (September 2007).

Chapter 8

Crowd dynamic loading on footbridges

James Brownjohn, Stana Zivanovic & Aleksandar Pavic
University of Sheffield, Sheffield, UK

SUMMARY

Design codes and guides concerned with vibration serviceability of footbridges have, in their present form, origins in research from the 1970s and until very recent innovations (which are still informative rather than normative) have used either frequency control or provision for response due to a single pedestrian. Clearly footbridges are expected to be fit for purpose when carrying multiple pedestrians, even large numbers or 'crowds' at special events and there have been several high-profile examples of footbridges that have failed to perform satisfactorily in these circumstances. These failures have highlighted shortcomings in current codes and design guides and led to a significant research effort to develop updated rational design guidance covering cases not just of single pedestrians, but also scenarios of multiple pedestrians, and effects for response in both vertical and lateral directions.

Proposed approaches for single pedestrian time-domain deterministic predictions include higher harmonics of the pedestrian forcing function and lateral forcing at the sub-harmonic of the pacing rate. There is also recognition that both loading levels and comfort thresholds for pedestrians are well described by probability density functions, and this naturally leads to probabilistic, risk-based approaches. Such approaches are well suited for the case of multiple pedestrians, in which case frequency domain approaches using random vibration theory may be applied.

Further development of such approaches requires significant levels of research on loads from pedestrians individually, in groups and in crowds. Numerous case studies have shown human structure interaction, and this has to be accounted for, assuming even that a reliable numerical model of the unloaded structure is available. Finally, a better understanding of comfort thresholds, their variability and their dependence on circumstance is required.

This chapter reviews the current knowledge about modelling walking-induced dynamic loading and pedestrian's interaction with both each other and the movement of the structure. It starts with the single pedestrian models (a necessary prerequisite), and then leads to reviewing different approaches to modelling streams of pedestrians and crowds. The models described include those based on multiplication factors for single person, Monte Carlo traffic simulations and frequency domain approaches. As for the interaction, mainly synchronous lateral excitation is addressed. Modelling and testing the footbridges are beyond the scope of this chapter, as is the evaluation of acceptable vibration levels for pedestrian receivers.

Finally, case studies of measured lateral and vertical responses are presented with some parallel simulation exercises using the various proposed approaches.

8.1   INTRODUCTION: PEDESTRIAN EXCITATION OF FOOTBRIDGES

With modern structural design, structural ultimate limit state failure of a footbridge is nowadays an extremely rare event, and historical examples of failed bridges described in [1] have all been ascribed to loading by crowds. Observations that failures were more likely when pedestrians (not limited to soldiers) walked 'in step' led to Victorian era classical (enforced) guidance to avoid synchronisation.

Given such a condition is avoided, footbridges still need to be fit for purpose. While the purpose is superficially obvious, all typical and possible loading scenarios need to be considered at the design stage. Clearly, a footbridge is not fit for purpose if a pedestrian is uncomfortable or frightened to use it under 'normal' conditions because it vibrates too much. Despite the condition of only a single pedestrian on a footbridge being far from the 'normal' condition in the majority of cases, especially in busy city centres, it remains the scenario of almost every design guide and code addressing the issue of pedestrian comfort. It also leads naturally into proposed approaches for simulating more typical conditions of multiple pedestrians ranging from independent walkers to small groups, steady streams and ultimately crowds packed on a bridge. The issue of which scenario to design for is addressed in [2] via a risk matrix of high probability/low severity (i.e. the normal event) to low probability/high severity events.

8.2   REVIEW OF KEY MODELS FOR SINGLE PEDESTRIAN

Numerous models of walking excitation have been proposed [3,4] and usually derive from the assumption that walking is a perfectly periodic activity. The simplest such model that uses a single sinusoidal function is found in current British [5] and Canadian [6] codes while vibration design guides primarily for floors [7,8,9] provide for up to four harmonics. This means that actual forces due to continuous walking can be re-created or synthesised, by adding a sequence of identical single footfall traces, temporally displaced by integer multiples of an exact footfall interval. The individual footfall forces, from which the harmonic amplitude values are usually obtained (from measurements of force due to walking over a 'neutral' force plate that is supposed not to affect the walking pattern) and the resulting synthetic time histories are assumed to be periodic. In a time domain analysis procedure that uses such a synthetic walking time series, the structure is assumed to be linearly elastic and, using modal decomposition, the response of each vibration mode can be analysed separately using harmonic loads modulated by the appropriate mode shape to account for the moving load [10].

Using Fourier analysis, any periodic and continuous signal $x(t)$ repeating itself at intervals T can be represented by an infinite ($m \rightarrow \infty$) sum of terms:

$$x(t) = a_0 + \sum_{n=1}^{m} a_n \cdot \sin(2\pi t n / T + \phi_n) \tag{1}$$

Hence, if we take a sample of duration T [s] from a synthetic time series for walking at a pacing rate of $\bar{f} = 1/T$ footfalls per second [Hz], the walking forcing function can be decomposed, via equation [1] into:

- a mean value a_0, equivalent to body weight, in the case when $x(t)$ presents a force induced by a pedestrian,
- a fundamental component ($n = 1$) at pacing rate $\bar{f}$ and,
- a sequence of $m - 1$ higher harmonics, where the value of m depends on which model is used. Some models even use up to the sixth harmonic [11].

The ϕ_n are phase angles which have been found to vary considerably between various measurements for higher harmonics [12] indicating the need to consider their random nature when generating dynamic forces due to human walking. The values of normalised coefficients $G_n = a_n/a_0$ are known as 'dynamic load factors' (DLFs) and depend on walking speed and individual human characteristics. Their values obtained from different sources are summarised in [13]. In one of the most comprehensive studies of DLFs [14], wide variability was found in DLFs generated by 40 test subjects, in numerous walking tests. This was due to differences between subjects participating in the tests (inter-subject variability) and also between different tests with the same test subject (intra-subject variability). This variability is described by a frequency-dependent mean and coefficient of variation for DLFs [15] that is now formally adopted in design guidance for walking induced vibrations in floors in the UK [9]. The mean DLFs defined in this way give rise to deterministic predictions of response to walking and form the basis for the most popular candidate models for crowd loading.

An extension of single person model to multi-harmonic approach is presented in [16]. The frequency-domain forcing content up to the fifth harmonic was fitted to treadmill data including all the spectral lines between the harmonic frequencies, as well as the sub-harmonics due to unbalanced gait. The model was then transformed to the time-domain assuming random phases between the spectral lines. Due to its representation of the forcing frequency content up to approximately 10 Hz, this model could be used for prediction of multi-mode response.

For lateral response, again for a single pedestrian, numerous values of lateral DLFs have been proposed for both walking on a rigid [17,18] and laterally moving surface [19]. In the former case the DLF is dependent on walking frequency for the first harmonic only (higher harmonics are not dependent on walking frequency) while in the latter case the first, and the second, harmonics are a function not only of frequency but also of acceleration.

8.3   PROVISIONS FOR CROWD LOADS IN LITERATURE AND CODES

Dictionary definitions of a crowd usually refer to a gathering of a large number of people, in close proximity. Between single pedestrians and a crowd, other design scenarios are light to heavy traffic of pedestrians, groups of people walking together and sports events such as marathons. There is very limited codified provision for dynamic loads on footbridges in any of these situations, the exceptions being French (Setra) guidance [20] and the Eurocode UK National Annex (EC1 NA) [21] using a multiplication factor applied to single pedestrian response and the limited British guidance resulting from the London Millennium Bridge experience and based on the obsolete frequency-tuning method. Inevitably, particularly in this millennium, most of the results are still in the research domain waiting to be more widely verified and codified.

8.3.1   Multiplication factors

Early research on crowd loading by Matsumoto et al. [22] suggested estimation of vibration response to normal pedestrian traffic by multiplying the response to a single person exciting the resonance of the footbridge by a factor $\sqrt{N} = \sqrt{\lambda T_0}$, where N is number of people on the bridge at any time instant, λ and T_0 are mean arrival rate, expressed as the number of pedestrian per second, and the time needed to cross the bridge expressed in seconds, respectively. This dependence was derived by summing responses of individual pedestrians arriving on the bridge according to the Poisson distribution and inducing the walking forces with equal (resonant) frequencies and random phases. The derived factor divided by total number of people N can be interpreted as a synchronisation factor, i.e. the portion of people in the crowd that, by chance, walk in step while the effect of the rest of the crowd can be neglected. On the other hand Wheeler [23] conducted Monte Carlo simulations for different crowd sizes considering the random nature of arrival time, body weight and step frequency. He argued that the increase in footbridge response to crowds (comparing with the single person response) is significant only for footbridges having natural frequency close to the mean pedestrians' pacing rate. Both studies [23,22] were of bridges vibrating in the vertical direction only. However, an attempt to make use of the Matsumoto formula on the laterally vibrating T-footbridge in Tokyo was made by Fujino et al. in order to estimate the vibration response [24]. It was found that the given relationship significantly underestimates the vibration response measured and it was argued that this was due to feedback effect, i.e. due to people adapting their movement to that of the laterally vibrating surface in an attempt to preserve their body balance. The authors concluded that for the T-bridge (accommodating as many as 2000 people at a time) instead of Matsumoto's synchronisation factor of $\sqrt{2000}/2000 = 0.022$ an almost nine times larger value of 0.2 should be accounted for in the vibration prediction. Additionally the Matsumoto formula could not predict the vibration level on the Millennium Bridge [25], suggesting that the approach is not viable for bridges prone to strong lateral vibration due to synchronised crowd.

The multiplication factor approach was also adopted in the French Setra guideline [20]. The guideline distinguishes between four classes of footbridges depending on traffic level. These classes are:

Class I: very dense crowd (1 ped/m²),
Class II: dense crowd (0.8 ped/m²),
Class III: sparse crowd (0.5 ped/m²), and
Class IV: seldom used footbridges.

Depending on natural frequency of the relevant vibration mode and footbridge class, an equivalent number of people for different combinations of classes and frequency ranges is defined (Table 8.1), where ζ is the modal damping ratio of the relevant vibration mode. The same reasoning is applied for the longitudinal direction, while for the lateral direction the walking frequency ranges are divided by two.

Multiplying the equivalent number of people from Table 1 by another factor $\psi \in [0,1]$, defining the risk of resonance that depends on the relevant mode shape frequency, gives a total equivalent number of people. However, differently from the previous approaches, this number should be uniformly distributed along the bridge length L and 'modal number of people' should then be calculated taking into account the (absolute value of the) shape $\phi(x)$ of the relevant vibration mode. Therefore, the multiplication factor in the classical sense (i.e. the one that should multiply the force induced by a single person) could be obtained once the equivalent number of people is multiplied by $\psi(\int_0^L |\phi(x)| dx)/L$. Finally, the multiplication factor is applied to the resonant response due to a stationary single person force. This is different from Matsumoto's approach which uses peak response generated during single person crossing (meaning that limited duration and mode shape are taken into account in the single person model).

The two expressions for the equivalent number of people defined in Table 8.1 are empirical results obtained using Monte Carlo simulations. In the case of Class I it was assumed that all pedestrians walk at the same frequency, but the phases are

Table 8.1 Equivalent number of people according to [20].

Footbridge class	Natural frequency range (Hz)			
	1.7–2.1	*1.0–1.7 or 2.1–2.6*	*2.6–5.0*	*<1.0 or >5.0*
IV	No calculation	No calculation	No calculation	No calculation
III	$10.8\sqrt{N\zeta}$	No calculation	No calculation	No calculation
II	$10.8\sqrt{N\zeta}$	$10.8\sqrt{N\zeta}$	$10.8\sqrt{N\zeta}$	No calculation
I	$1.85\sqrt{N}$	$1.85\sqrt{N}$	$1.85\sqrt{N}$	No calculation

random (uniform distribution), while for Class II/III, apart from random phases, the randomness in the pacing frequencies is also assumed (Gaussian distribution). As would be expected, the DLF in the single person model used in the multiplication depends on the excitation and vibration direction, being 0.4, 0.2 and 0.05 for the first harmonic of the vertical, longitudinal and lateral force, respectively, assuming a pedestrian's weight of 700 N.

Tentative pedestrian design load models defined as DLM1, DLM2 and DLM3 were originally proposed for Eurocode 1 and catered for a single person loading, small groups of 8–15 pedestrians and steady streams, respectively. These models were not successfully calibrated so were not implemented [26]. Nevertheless, these models are published in design guidance from FIB [27]. DLM1, which is the basis for DLM2 and DLM3, is defined as a sinusoidal force, in horizontal lateral or vertical directions, at a frequency matching the natural frequency of the relevant vibration mode. Presumably, the intended amplitude of the force in the vertical direction is 280 N, which is the same as in Setra guidance [20] (although there is contradictory information in FIB document referring to both 280 N and 180 N) while the lateral value of 70 N is twice that defined in [20]. Although the speed of a walking pedestrian is suggested to be 0.9 times the walking frequency, the final model is defined with a stationary force applied at the (maximum) nodal point of the mode shape. DLM2 is DLM1 multiplied by the synchronisation factor dependent on the natural frequency, having maximum value of 3 for natural frequency in the most probable walking frequency range (1.5–2.5 Hz). In this case an 800kg mass is required to be added to the nodal point, to account for pedestrians' influence on the frequency of the human-structure system. Lastly, the DLM3 model assumes that the bridge is uniformly loaded by resonant dynamic force due to 0.6 pedestrians/m^2 multiplied by frequency dependent synchronisation factor (with maximum value of 0.3 and minimum of 0.05) and a reduction factor of 0.75 that takes into account limited duration of the individual pedestrian's force. Also, to account for human structure interaction (HSI) additional mass of 40 kg/m^2 is to be added. Finally, the force is to be applied only on sections of footbridge having the same sign of the mode shape, to simulate the worst case scenario.

Finally, EC1 NA [21], which is designated as informative so that the use of alternative approaches is allowed, suggests that it is difficult to define a unifying approach for modelling pedestrian loading. Similarly to Setra guideline, EC1 NA differs between four classes (A, B, C and D) of footbridges depending on the traffic expected. The classes range from class A which is a seldom used, rural bridge to class D which represents busy bridges used as primary access to public assembly structures, such as stadia. For each class, the number of people likely to walk and jog in a group is suggested, with group size increasing from minimum for class A (2 for walking, 0 for jogging) to maximum number of people for class D (16 for walking, 4 for jogging). The vertical response of the bridge is then calculated for each mode of interest by multiplying the response due to a single person walking at the resonant frequency by factor $\sqrt{1+\gamma(N-1)}$ where N is the number of people in the group while γ is a factor that takes into account the synchronisation effect between people in the group. The response is further multiplied by factor k that accounts for the excitation potential of relevant forcing harmonic and probability of walking at the resonant frequency.

Similarly, crowd loading is defined as a load per unit area w pulsating over the deck at the resonant frequency, and with the load sign matching the sign of the mode shape. This load is obtained by multiplying the resonant loading from a single person over area of interest by the factor:

$$1.8k\sqrt{\frac{\gamma N}{\lambda}},$$

where N is the number of people on the deck at the same time, factors γ and k are as already explained, while λ is a factor that reduces the number of effective pedestrians when only a part of the span contributes to the mode of interest. Accounting for net effect of several vibration modes could be done by calculating the 'vector sum' i.e. SRSS, to be on conservative side. More information about development of the model in NA can be found elsewhere [28].

8.3.2 Monte Carlo simulations and design spectra

Monte Carlo (MC) simulations seem to be a convenient way of predicting the footbridge response to pedestrian traffic when pedestrian movement is not spatially restricted. This is probably the most realistic design case for most footbridges. The approach consists of 'firing' the pedestrians along a bridge according to (usually) Poisson distribution, while the parameters such as walking frequency, pedestrian weight, speed and force amplitude could be generated from the appropriate probability distributions. These simulations inevitably make use of a single person model that has to feature all parameters taken as random variables in the simulations. The knowledge about distributions of these parameters within the pedestrian population is required. After Matsumoto's study, there was little interest in MC simulations until recently, when the valuable data needed for the simulations were acquired by numerous researchers for walking [29,30,31,18,16,32,33] or running in general [34] or during a marathon event [35].

Based on MC simulation of response of a 'standard bridge' to a 'standard population of people' design spectra representing the peak response of the bridge in a single vibration mode as a function of natural frequency were constructed [36,37]. These design spectra are similar to those used by the earthquake engineering community where peak modal response values are presented as a function of mode frequency for a specific input. The 'square root sum of squares' (SRSS) approach is used to sum the individual modal responses calculated in this way.

8.3.3 Frequency domain approach using power spectral density

The periodic model of the walking-induced force lends itself to deterministic time domain analysis. On the other hand, the dispersed energy around and between harmonic frequencies, which can be seen for real continuous walking, is more suited to the techniques of stochastic analysis in frequency domain, where the narrow-band walking forces are presented as power spectral densities (PSDs). To illustrate the dispersion of energy around the harmonic frequencies, Figure 8.1 (left) shows a real walking time history measured on an instrumented treadmill and its periodic synthetic

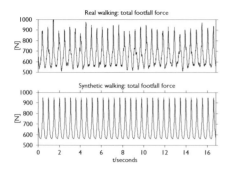
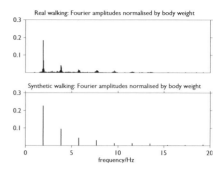

Figure 8.1 Time-histories (left) and Fourier amplitude (line) spectra (right) of real and synthetic forces corresponding to the pacing rate of 1.91 Hz.

counterpart obtained by averaging the time histories of individual steps. Figure 8.1 (right) shows Fourier amplitude (line) spectra of both time histories. It is clear that in real walking energy is spread around the spectral lines of dominant harmonics at pacing frequencies and their integer multiples due to imperfect walking and energy distribution around a harmonic line can be modelled as a PSD.

When applied with the Gaussian form of probability distribution of mean pacing rates, first described in [22] and confirmed by others later [30, 38, 33], application of PSDs leads to a framework for more realistic frequency domain force representation for individuals and groups of pedestrians.

The dependences on $\sqrt{N}$ and $\sqrt{N\zeta}$ found empirically in the Setra study [20] are results from random vibration theory applied in the context of wind engineering [39] as well as for pedestrian loads modelled as random processes. This approach has been demonstrated in [40] and discussed in a later section in the context of a 140 m span footbridge. The analogy with wind engineering also suggests the possibility of modelling correlation between pedestrians at different locations on a structure via a distance-dependent coherence function.

Butz [41] suggests deriving PSD functions (and therefore RMS values) of vibration response using Monte Carlo simulations that account for probability distributions of pedestrian mass, step frequency, walking speed, forcing amplitude and pedestrian arrival times in a stream of pedestrians. Taking into account interdependence of some parameters (such as walking speed, step frequency and stream density), she expressed the expected response as a function of empirically defined parameters. This approach is promising, but it requires expressing all relevant parameters in a form convenient to use in design.

8.4   SYNCHRONISATION AND LOCK-IN

Synchronisation in the context of pedestrian dynamic loads is taken to mean the tendency among pedestrians to walk at the same pacing rate as each other, while lock-in describes the tendency to synchronise pacing rate with strong vibration of the bridge.

Lock-in may or may not occur on a laterally moving surface [42] depending on an individual's style of walking. However, if it occurs it may trigger a mechanism for synchronisation.

8.4.1   Synchronous lateral excitation (SLE)

As a result of the vibration serviceability failure of the London Millennium Bridge in June, 2000, a considerable research effort went into investigating the problem and finding a remedy. Studies on lateral forces on a harmonically driven swaying walkway [43] were carried out to determine the nature of the lateral loading by pedestrians, and full-scale measurements [44,45,46] were made to identify the dynamic properties in terms of vertical and lateral vibration modes. The semi-empirical model, presented by Arup [47] to explain the observed behaviour, proposed the force exerted by a pedestrian to maintain balance being directly proportional to lateral velocity, in effect a form of negative damping. The net damping, naturally existing in the structure occupied by people, would vanish as pedestrians reached a critical number, dependent on the mass, natural frequency and structural damping of the empty bridge. In pedestrian tests Arup validated this approach [48] and showed that SLE could occur in any bridge subject to a few conditions. The study provided evidence that the effect had been observed in the past on a number of bridges. In fact, the effect was mentioned in a widely regarded text on structural dynamics [3] and noted in Fujino's T-bridge paper [24], although the Arup treatment was the most comprehensive and first to offer an interpretation and mathematical model of the problem as well as its solution.

The idea is that by chance a few pedestrians walk coherently and the resulting lateral response is sensed by more and more of the available crowd of pedestrians. Once a certain size of crowd (or, it is also argued, a certain level of vibration) is reached, and there is a very conscious adjustment of gait to maintain balance, providing positive feedback, the response continues to grow. The observations have since been repeated on a number of footbridges but as yet there is no convincing explanation of the complete process.

The Arup formula for the critical number of pedestrians in a crowd for SLE to occur is

$$N = \frac{8\pi f_r \zeta_r m_r}{k} \tag{2}$$

where k is a constant estimated as 300 Ns/m and f_r, ζ_r, m_r are, respectively, modal frequency, damping and mass for the affected lateral vibration mode. Subsequent to the sequence of events culminating in publication of the Royal Academy of Engineering Paper [47], there has been enhanced interest in the problem, even as far as a review paper [49] that notes the limitations of the various theories, including Arup's original positive feedback mechanism.

In summary, [49] considers three forms of model to describe the phenomena of SLE: direct resonance, dynamic interaction and internal resonance and adds a new model.

Direct resonance [24] uses a sinusoidal forcing function due to a proportion of the crowd, in resonance with the relevant lateral mode of the bridge. This works where synchronisation is achieved and the pacing rate is twice the lateral mode frequency.

Dynamic interaction is the most popular variant, where the dynamics of a pedestrian become linked to the dynamics of the bridge. The Arup model is of this form, as are variants due to Roberts [50], Nakamura [51] and Newland [52] who propose various mechanisms for the interaction.

A third form is internal resonance, which also includes parametric excitation. An example of parametric excitation is lateral vibrations of a cable whose axial tension reaches a maximum every half cycle of the cable oscillation, and vice-versa. This provides the possibility that pacing at, say 2 Hz, with lateral component at 1 Hz can force response in a mode at 0.5 Hz, hence providing an alternate (to the dynamic interaction approach) explanation for excitation of the Millennium Bridge fundamental mode at 0.5 Hz.

One condition for parametric excitation would be when a mainly-lateral mode frequency occurs at half the frequency of a mainly-vertical or torsional mode, due to the minor response component (vertical and lateral respectively). This type of model has been proposed in [53], in relation to the Solferino Bridge.

The new model from [49] works in both parametric and interaction modes, covering cases where modes are forces at 1 and 0.5 times the pacing rate.

Venuti et al. [54] rightly argue that the walking people have to be modelled as part of a complex dynamic system rather than as an external force only. They suggest a non-linear model for human-structure dynamic interaction while walking in crowd. They model the crowd dynamics using the mass conservation equation (linking crowd density and walking speed) while the structure dynamics are modelled using the equation of motion for a SDOF system representing a single vibration mode. The feedback effect is accounted for through the force term that is a function of deck velocity, crowd density and walking speed. There are two other studies worth mentioning. Strogatz et al. [55] use the parallel of biological rhythms being affected by weak periodic stimuli. In this case the rhythm is pacing and the weak stimulus is the response of the bridge to random forcing. The result is a critical number for 'wobbling' and synchrony to occur simultaneously. Barker [56] provides the mechanism for the weak stimulus. He shows that even with pacing at frequencies other than twice the lateral mode frequency, the work done in the lateral direction by a pedestrian as a result of the weight resolved along an inclined leg is an asymmetric function. Hence even uncorrelated pedestrians put energy into the lateral mode. This is a method the authors believe holds most promise for explaining the phenomenon and is recently further developed by Macdonald [57].

8.4.2　Vertical lock in and Scruton Number

Bachmann & Ammann [3] reported that vibration amplitude in the vertical direction higher than 10–20 mm probably forces a test subject to change their step, and adjust to the motion of the deck. However, based on tests on Solferino bridge [20] it was suggested that no lock-in in the vertical direction seems probable, due to pedestrians being disturbed by strong vibrations in this direction and not being able to maintain the walking at the resonant frequency. This supports the results of a previous study dealing with effects of vibration during single pedestrian tests on three footbridges [58]. Also, as a demonstration of human-structure interaction (HSI) this is consistent with a view that there is some form of increased damping due to the pedestrians walking [17, 38, 59] caused by footfalls acting as 'active damping forces' preventing the structure from moving freely. This is probably due to human's inability to synchronise their motion with excessively moving structure in the vertical direction. When this motion is not excessive but still perceptible, the synchronisation is possible, but there is a considerable drop of the contact force, an effect similar to the shaker drop-off force when performing step-sine modal testing around resonance. This is clearly demonstrated in [60] for jumping. One theory is that since this vertical synchronisation is not crucial for preserving the balance and stability, as opposed to the human reaction to the lateral movement, quite often in the vertical direction humans act as 'active dampers' supplying force which opposes the structural motions, thus effectively dampening it. A recent experimental study supports this claim [61].

Contrary to the observations for jumping in [59], there has been a suggestion that synchronisation and unbounded response can actually occur for walking in the vertical direction. This is yet to be clearly proven, and has to be set against the proven positive damping capability of pedestrians for vertical response. However, an equivalent of the Scruton number used in wind engineering has been proposed [52, 62] to provide a form of reduced damping that can be applied to footbridges to check both vertical and lateral stability. The form used for footbridges would be

$$S_{cp} = 2\zeta M / m \qquad (3)$$

where S_{cp} is pedestrian Scruton number, ζ is modal damping ratio, M is bridge mass per unit length, and m is pedestrian mass per unit length. A higher value is better, and for both vertical and lateral responses S_{cp} should exceed approximately 0.27. This procedure is not widely used (for lateral response), compared to the Arup formula.

Using a similar (Scruton number) approach, the EC1 NA [21] suggests designing bridges to avoid possibility of unstable response in the lateral direction. The stability has to be checked if there are lateral modes with natural frequency below 1.5 Hz and it depends on the relevant natural frequency of the lateral mode and the pedestrian mass-damping parameter D defined as:

$$D = \frac{m_{\text{bridge}}\zeta}{m_{\text{ped}}} \qquad (4)$$

where m_{bridge}, m_{ped} and ζ are the bridge mass per unit length, the pedestrian mass per unit length and damping ratio of the lateral mode in question. If the combination of the natural frequency and parameter D belongs to a stable part of the diagram defined in the code based on the experimental data available, then the problems with its lateral instability due to crowd loading are not expected. The reliability of the provisions defined in the EC1 NA is yet to be tested on real-life structures.

8.5   CASE STUDIES IN MEASUREMENT UNDER CROWD DYNAMIC LOADING

Since the problem experienced by the London Millennium Bridge, with enhanced interest in the topic as well as increased concerns and awareness by footbridge stakeholders, there have been several experimental studies published. The emphasis in these studies has naturally been on lateral performance, but there have been some useful results for vertical loading due to crowds, which predate the Millennium Bridge problems. These are essentially distinct problems and because SLE is essentially an instability problem, it requires more drastic measures than the vertical liveliness.

8.5.1   References to experimental studies of crowd loading on full-scale structures

A historical survey of crowd-induced failures of footbridges [1] shows four examples involving cavalry or soldiers crossing suspension bridges, which led to the requirement for soldiers to break step when crossing bridges. Significant vibration levels are expected when crossing suspension bridges without stiffening trusses, but such response would not be acceptable in conventional apparently rigid structures and pedestrians would avoid using them.

Hence followed the development and adoption of vibration serviceability standards worldwide, with 'failure' now being interpreted in terms of poor serviceability i.e. 'not being fit for purpose'. Without any evidence of structural collapse, reports of such failures are less evident except in spectacular cases such as Solferino and London Millennium Bridges, where in each case it was lateral vibration levels that were unacceptable.

As previously mentioned, these two cases were in fact not the first occasions that surprisingly large lateral vibration had been observed, nor the last. As well as the Tokyo T-bridge [24], major highway bridges including the Bosporus Bridge [1] and Auckland Harbour bridge experienced significant lateral response during occasions of heavy crowding and there is anecdotal evidence (from press and internet blogs) for the same effect on the Brooklyn Bridge during blackouts in New York in 2003. More recently, the effect was observed in the Clifton suspension bridge in Bristol, UK [63] during a festival.

Crowd loading studies on Millennium and Solferino Bridges are reported in [47] and [64] respectively, and since then many operators of new bridges, mainly in Europe, have commissioned crowd loading tests. Some of these exercises that have reached public domain include Changi Mezzanine Bridge in Singapore [65], Coimbra footbridge [66] and the Tokyo T-bridge [67]. Some of these cases of SLE are discussed in the next section.

There have also been several research exercises in which responses of bridges under heavy in-service pedestrian use have been studied e.g. Podgorica Bridge,

Montenegro [38, 68], Western Approach Footbridge, Plymouth [69], Singapore Polytechnic walkway [70] and Tanjong Rhu footbridge, Singapore [71].

8.5.2   Published case studies for footbridges exhibiting SLE

Due to increased awareness in this millennium, researchers have taken greater care to document cases of SLE, while operators have been motivated to allow such comprehensive studies before putting a bridge in service. Hence five example studies can be summarised here.

8.5.2.1   London Millennium Bridge

The story of the London Millennium Bridge (LMB), Figure 8.2, is described through a number of papers, including a comprehensive overview in [72]. The design and construction are also described in [25, 73, 47, 48]. This suspension bridge has three spans of 81 m (north), 144 m (centre) and 108 m (south). The suspension cables are two groups of four 120 mm locked coil wire ropes and the centre span sag is an unusually small 2.3 m (conventional suspension bridges have sag-to-span ratios in the 1:8–1:10 range) leading to a large tension in cables of 22.5 MN to carry

Figure 8.2 London Millennium Bridge, showing vertical TMDs and viscous dampers installed at crossbeams and dampers connecting to the tower.

the 2000 kg/m dead load of the bridge. The deck comprises extruded aluminium box sections spanning between edge tubes, and these are supported on fabricated aluminium boxes suspended from the cables. Wind loading studies were aided by wind tunnel testing and pedestrian loading was based on the BS5400 [5] single 'bad man' provision for all susceptible vertical vibration modes but using a higher input force than recommended, with a maximum vertical peak acceleration response of 0.19 m/s^2 predicted. The response in lateral modes was also predicted and found to be acceptable.

Between 80 and 100 thousand people crossed the bridge during the opening day on 10th June 2000. Strong lateral vibrations were observed, mainly on the south span at a frequency of around 0.8 Hz and also in the centre span at frequencies of 0.5 Hz and 0.9 Hz. Maximum peak acceleration values were believed to be up to 2.5 m/s^2. At this level, pedestrians had difficulty in walking without support. Fewer pedestrians were allowed on the bridge on 11th June, and on 12th June it was decided to close the bridge for investigations.

A programme of research followed the closing [47]. Laboratory tests were commissioned at University of Southampton and at Imperial College London [43] to determine the nature of the lateral force induced by a pedestrian and the dependence on lateral motion. Full-scale walking tests were also conducted on the bridge using up to 275 pedestrians circulating on the bridge on the critical spans, increasing the number of pedestrians until large amplitude response was observed. The spectacular result of these tests is shown in [73], and analysis of the proposed mechanism for SLE, in form of equation [2], is presented in two papers [48, 47]. While the proposed approach, based on assumption of forcing frequency being close to a natural frequency of the bridge, is logical for excited lateral modes with frequency close to 1 Hz, it seems less viable for the mode at 0.5 Hz. However, for this mode alternative explanations that the frequency of human walking reduces in crowd and that people exhibited 'meandering' patterns in their walking were offered [25].

Successful FRF-based modal testing of the Bridge carried out by the University of Sheffield and a team of sub-contractors took place in December 2000 during which characteristics of lateral and vertical vibration modes were identified [44, 45]. By using shaker testing in the main span, lateral modes of the centre span were found at 0.47 Hz with modal mass around 100 tonnes, and at 0.95 Hz with modal mass around 150 tonnes. The testing also evaluated the performance of proposed remedial measures, including tuned mass dampers for vertical and horizontal directions and discrete damping elements (viscous dampers) for horizontal modes.

A large number of vertical TMDs were installed, and are clearly visible on the cross beams of the bridge but the retrofit [48, 44, 45] to control lateral vibration primarily relies on dampers connecting the deck to the towers or to ground and positioned across diagonal bracing elements under the deck floor. Modal damping levels are now considerably higher than the original 0.5% to 0.6% and there is no longer any realistic prospect of SLE happening again.

8.5.2.2 Solferino Bridge

Compared to London Millennium Bridge, relatively little material is available in public domain for this bridge. The significant public domain research article is in French [64], but many useful results are also available in the Setra guidelines [20] and the modal testing results are used in the study by Blekerman [53].

Originally, the fundamental lateral mode for Solferino Bridge occurred at 0.81 Hz and pedestrian tests demonstrated that with a sufficiently high acceleration response, lock in would occur, although the effect was not as spectacular as with London Millennium Bridge. The Solferino bridge walking tests demonstrated that rather than being dependent on a critical number of pedestrians, SLE would initiate when lateral acceleration response exceeded a threshold, estimated at around 0.15 m/s^2, with one test clearly demonstrating the effect using 229 randomly walking pedestrians. Also, the maximum proportion of synchronised people during the tests was estimated to be about 60%.

The retrofit comprised installation of numerous 2500 kg TMDs for both vertical/torsional and horizontal vibrations, after which the first lateral mode dropped to 0.71 Hz, with damping increased from about 0.5% to approximately 2.6%. The modal mass for this mode was estimated as 260 tonnes using forced vibration testing after the retrofit.

8.5.2.3 Changi Mezzanine Bridge

Changi Mezzanine Bridge (CMB) [59] was opened in 2002 and provides a mezzanine level underground walkway between two terminals at Singapore's Changi Airport. The bridge, shown in Figure 8.3 (left) has a main span of 140 m in the form of a shallow arch. The total mass of the bridge is estimated as 1300 tonnes corresponding to 6500 kg/m (over three times the value for LMB) and total deck area available for pedestrians is approximately 840 m^2.

The chronology of CMB and LMB are closely linked as shown in Table 8.2. By the time construction of CMB was well underway, results from the LMB study were becoming available and analytical and experimental studies were commissioned to investigate possible excessive vibration due to crowd loading.

Preliminary studies identified two lively vibration modes, the first symmetric lateral vibration mode (LS1) at approximately 0.9 Hz and the first symmetric torsional vibration mode (TS1) at approximately 1.64 Hz, both with damping ratios of approximately 0.4%. Modal testing using shakers was used to map out mode shapes for these and other modes, and modal masses for these critical modes were also estimated at approximately 450 tonnes (LS1) and 145 tonnes (TS1).

Immediately following the modal testing, a program of walking tests with up to 150 pedestrians was used to investigate the possibility of SLE. Figure 8.4 shows the result of this test for both modes.

The figure shows a continued build-up of lateral vibration response in mode LS1 even after the maximum number of pedestrians were circulating on the main span of the bridge. The maximum vibration levels were around ten times smaller than for LMB i.e. much less alarming, but still uncomfortable. Further tests showed that SLE could occur with smaller numbers of pedestrians, even as few as 100.

The same procedure developed by Arup [48] to recover the pedestrian forces was applied and the value of the coefficient k in the SLE formula was determined to be 188 Ns/m before accounting for mode shape effects, after which the value would be very close to the value of 300 Ns/m for LMB.

Table 8.2 Chronology of CMB.

CMB Design awarded to Arup (New York) and SOM	4th July 1997
London Millennium Bridge (LMB) opened	10th June 2000
LMB closed	12th June 2000
CMB Construction begins	June 2000
Report on CMB vibration serviceability issued	December 2000
Walking tests on LMB to validate Arup theory on SLE	19th December 2000
CMB skeleton frame completed	June 2001
CMB Prototype test	31st January 2002
CMB walking tests	1st February 2002
CMB partially open to public	8th February 2002
TMD installed and tested (bridge still not fully open)	July 2003
Changi Airport Terminal 3 opened	9th January 2008

Figure 8.3 Changi Mezzanine Bridge, Singapore (left) and pendulum TMD to control lateral mode (right).

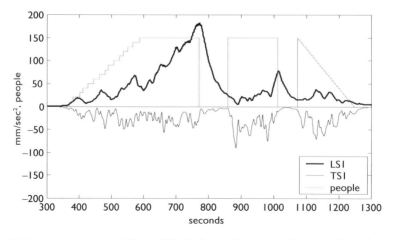

Figure 8.4 Response in modes LS1 and TS1 during walking test with up to 150 pedestrians.

Following the testing, a pair of TMDs with total mass of 1000 kg was installed at the centre span, after which the damping of the LS1 mode was shown to have been increased from approximately 0.4% to approximately 1.6%. The TMD is shown in Figure 8.3 (right).

No provision was made for control of vertical response. Figure 8.4 shows vertical response less than 100 mm/s^2, which is below the response predicted for the critical mode (TS1) using BS5400 [5] approach. While initial predictions were of very large response for this mode, clearly there was not sufficient synchronisation between pedestrians walking at an appropriate pacing rate, moreover there is evidence of increased damping ratio (and frequency changes) for this mode due to the presence of pedestrians, which was not observed in any way for lateral response. In other words, vertical response is partially mitigated by the pedestrians themselves as a form of HSI but there is no HSI effect on lateral response apart from the supposed SLE mechanism. The result for vertical response mitigation is in line with recent findings for crowd excitation of stadium grandstands [74] and effects observed on LMB [17] and Podgorica bridge [38].

A corollary of this result is that there was no evidence of synchronisation of pedestrians at a pacing rate twice the LS1 frequency. This is in line with the findings for the Clifton Suspension Bridge [63], described in section 8.5.2.5.

8.5.2.4   Coimbra footbridge

The footbridge over the Mondego river in Coimbra, Portugal [6], shown in Figure 8.5 (left), has a total length of 275 m, comprising two steel half arches and a 110 m composite centre span. The unusual feature of the bridge is the lateral anti-symmetry due to offsetting the two halves of the bridge which meet in a 8 m × 8 m square at midspan. Preliminary studies with the involvement of ViBest (http://www.fe.up.pt/vibest) in Porto suggested vulnerability to lateral or vertical pedestrian-induced vibrations, so a passive control system based on several lateral and vertical TMDs was proposed.

Ambient vibration measurements before completion and installation of all pavement and handrail finishes validated finite element modelling of the bridge used for the simulations and helped to fine tune values for foundation stiffness. A lateral mode was identified at 0.9 Hz. This mode, with damping 0.5–0.6%, was confirmed in the free decay tests. Based on these values and an estimated modal mass of approximately 200 tonnes from the validated FE model (and subsequently checked by forced vibration testing), the Arup formula predicted that 73 pedestrians would trigger SLE.

Figure 8.6 (left) shows that during walking tests with students, a significant increase in lateral response (leading to a clearly visible sway) resulted from exceeding just this number of pedestrians. Working back from an acceptable limit to pedestrian numbers led to a design of a system of six individual TMDs (Figure 8.5, right) each with mass of 2465 kg.

Following the installation, a forced vibration test was used to confirm the effectiveness of the system. The lateral mode frequency of the bridge dropped to 0.83 Hz with the TMDs inactive (and the same 0.55% damping), but with them active, two modes were found, 0.79 Hz (6.5% damping) and 0.88Hz (4% damping) in the case where sufficient force was applied to activate the dampers. It was observed that the dampers did not achieve their full design efficiency; this was also observed for the

Figure 8.5 Pedro and Ines footbridge, Coimbra (left), and TMD installation (right) (courtesy E. Caetano).

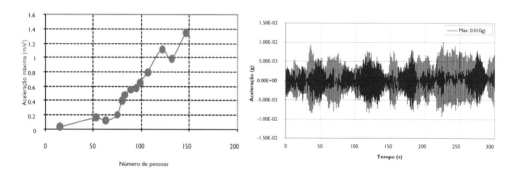

Figure 8.6 Relationship of response to pedestrian numbers during walking tests (left) and (right) response on inauguration day (courtesy E. Caetano).

much simpler installation at CMB. In any case, the performance of the bridge is now satisfactory, and the response observed on the inauguration day in November 2006 (Figure 8.6, right) is much reduced, and the frequency spectra show the response to comprise a strong mode at 0.81 Hz and a weak mode at 0.76 Hz.

8.5.2.5 *Clifton Suspension Bridge*

Clifton Suspension Bridge (CSB), Figure 8.7, was designed by Brunel and completed in 1864. Its main span is 214 m (between cable saddles), with side-spans of 60 m. The 9.5 m wide deck comprises a 6.1 m wide roadway and footways either side. Wrought iron chains are continuous from the anchorages, and the deck is carried via suspension rods.

In 2003 a comprehensive ambient vibration survey of the bridge [63] identified several vibration modes, the most significant of which are, from the point of view of pedestrian-induced response, second lateral mode L2 at 0.52 Hz with 0.6% damping

Figure 8.7 Clifton Suspension Bridge (courtesy JHG Macdonald).

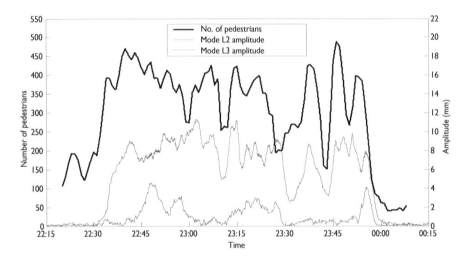

Figure 8.8 Lateral response of CSB at midspan and pedestrian numbers
(courtesy JHG Macdonald).

and mode L3 at 0.76 Hz with 0.7% damping. As no forcing function was used, no measured modal mass values are available, although the total mass of deck and chains between supports is estimated as 1150 tonnes.

Measurements of vertical and horizontal vibrations were made close to midspan of the bridge during a six-day period in 2003 that included the Bristol International Balloon Fiesta, when the bridge experienced heavy pedestrian traffic. Lateral displacement response measurements during the late evening of 7th August 2003 (Figure 8.8), obtained from double integration of acceleration signals, shows that modes L2 and L3 contributed to significant response levels (corresponding to 0.2 m/s²). The measurements show that strong vibrations were first initiated in the lower frequency mode L2, with strong response also occurring in the higher frequency mode L3.

There is no evidence of enhanced vertical response at twice these frequencies, instead the relatively broad response band from 1.4–2.2 Hz was excited. Applying the

Arup formula the critical number of pedestrians was calculated for each mode. These numbers were in the range 150 to 350 pedestrians, depending on damping values used, but in any case less than the actual number of pedestrians counted on the bridge during the strong response. The study, therefore, provides further evidence that SLE can occur with a critical number of pedestrians but does not support the traditional model for SLE, pointing instead to a mechanism such as proposed by Barker [56] which has been developed and partially validated by Macdonald [57].

8.5.3   Published case studies of lively response to vertical loads

Three examples from the authors' experience are presented here. For the last two examples, response simulations are presented for vertical excitation by the fundamental component of the pedestrian forcing function. The simulations use reliable (validated) finite element models of the bridges and a MATLAB-based simulation tool VSATs [75].

8.5.3.1   Singapore Polytechnic walkway

This 46 m long steel pedestrian bridge (Figure 8.9) is a steel space truss and frame system with the deck floor comprising timber planks with a layer of stone tiling. Only one end is fully supported, the other (the nearer in Figure 8.9) is connected to the adjacent structure by a narrow elastomeric expansion joint leaving a 15 m cantilever.

A combination of ambient vibration measurements and frequency response function estimation using a force plate and jumping pedestrian identified the first vertical mode with amplitude dependent natural frequency of about 5 Hz and 1.3% damping, and modal mass around 8500 kg. Using a force plate, the bridge was used to

Figure 8.9 Walkway at Singapore Polytechnic (left) and walking test (right) showing expansion joint at cantilever end.

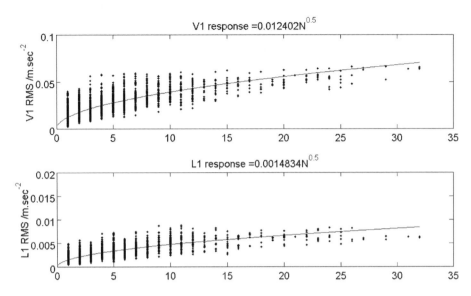

Figure 8.10 Correlation of response levels with pedestrian numbers N (horizontal axis),
ModeV1 Frequency is 5 Hz, Mode L1 frequency is 2 Hz.

study the ability of pedestrians to both supply and dissipate energy, thus controlling
the vibration levels. Additionally, during periods of heavy use, response levels were
monitored along with numbers of pedestrians using the bridge. These results are shown
in Figure 8.10, in which the square root dependence on N by forcing at the second or
third harmonic of walking is clear for both lateral and vertical modes/directions.

The single pedestrian vibration level predicted using BS5400 [5] is 0.105 m/s²
peak, a value which was frequently exceeded during the measurements (peak values
up to 0.3 m/s² were observed).

8.5.3.2   Podgorica Bridge, Montenegro

Podgorica footbridge is a steel box-girder bridge spanning 104 m over the Morača
River in Podgorica, Montenegro (Figure 8.11, left). The main span between inclined
columns is 78 m long, two side spans are 13 m each while deck width is 3 m. The
bridge vibrates perceptibly in the vertical direction under everyday human-induced
walking excitation.

To identify the lowest modes of vibration in both the vertical and horizontal
lateral directions, frequency response function (FRF) based modal testing (without
pedestrian traffic present) was employed. To reduce interruption of testing procedure by pedestrians crossing the bridge, the tests were conducted during night time,
starting at midnight. Pedestrians were allowed to cross the bridge during the breaks
between two successive tests.

Good quality FRFs in the range 1–9 Hz were obtained using a shaker with a chirp
excitation. The first vertical (1V) vibration mode, that is symmetric, with natural

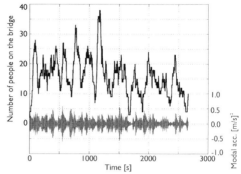

Figure 8.11 Podgorica footbridge (left) and modal acceleration and the corresponding number of people during a 45 min long test (right).

frequency of 2.04 Hz, modal mass of 58000 kg and extremely low damping of 0.22% was identified as the source of bridge liveliness. In a free decay test it was found that the damping increases to 0.26% for higher vibration magnitudes such as those generated by human-walking.

The Podgorica footbridge was also subject to vibration monitoring under normal pedestrian traffic on two different days. During total monitoring time of 4.5 hours, about 3000 people crossed the bridge, with an average number of 14 people present on the bridge at any time. A typical 45 min response in 1V mode measured at its nodal (midspan) point is shown in Figure 8.11(right) together with the total number of people present on the bridge at any time. The presence of moving people slightly reduced the vibration frequency from 2.04 to about 2.00 Hz. Modal response in mode 1V was almost the only contributor to the total response during monitoring since the other modes were hardly excited by pedestrian traffic.

The modal peak acceleration response of the bridge to single person excitation was up to 0.5 m/s^2 for free walking and up to 0.7 m/s^2 for metronome controlled walking, while during normal multi-person pedestrian traffic it went mainly up to 0.4 m/s^2, and quite rarely reached or slightly exceeded 0.6 m/s^2 (Figure 8.11, right). Maximum number of people on the bridge at a time was around 80 during pedestrian traffic monitoring tests, which corresponds to the density of 0.26 pedestrians/m^2. Also, during an interviewing session about 31% of pedestrians complained about vibration level perceived during footbridge crossing [75].

Based on 4.5 hours of traffic and vibration monitoring, the cumulative distribution of peak accelerations (per each vibration cycle) is shown in Figure 8.12 (left, solid line). The maximum acceleration in the same figure is 0.79 m/s^2, while 95% of peak values are below 0.35 m/s^2.

The Scruton number calculated according to equation [3] for this bridge of 260 tonnes and with, for example maximum of 80 pedestrians (weighting 75 kg each) uniformly distributed across the bridge is equal to 0.23. This is below the provisional limit of 0.27 given in [52], suggesting that the bridge is prone to vibration in the vertical direction, as observed.

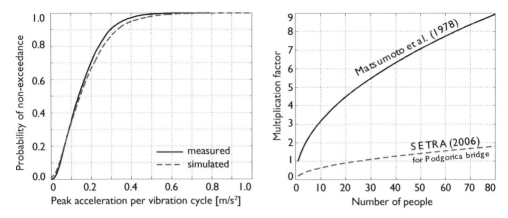

Figure 8.12 Cumulative distribution of the peak acceleration per vibration cycle as measured and calculated (left) and multiplication factor as a function of number of people on the Podgorica bridge (right).

8.5.3.2.1   Simulation using Matsumoto et al.'s (1978) model

If a single pedestrian, weighting 700 N, walks at footbridge natural frequency of 2.04 Hz with speed 1.5 m/s and induces a force with DLF = 0.4, i.e. with magnitude of 280 N (which are parameters defining a single person model in the Setra guide-line), then the peak acceleration generated is 0.45 m/s². Clearly, this acceleration level already covers around 98% of measured peak values (Figure 8.12, left, solid line). Multiplication factor $\sqrt{N}$ [22] as a function of number of people N is presented in Figure 8.12 (right, solid line). The predicted acceleration level for mean number of 14 pedestrians would be 1.68 m/s², while for the maximum of 80 people it would be 4.0 m/s², therefore significantly overestimating the measured values. The meas-urement results, therefore, support Wheeler's suggestion [23] that the single person response is a good enough estimate of multi-person traffic, even on this bridge that has natural frequency close to dominant pacing rate, as long as the single person is modelled with a realistic DLF dependent on the walking frequency. Had BS5400 [5] DLF of 0.257 been used in conjunction with proposed speed at 1.84 m/s, then the predicted response would be 0.26 m/s² only. This is a clear underestimation of the response to normal traffic.

8.5.3.2.2   Simulation using Setra model and NA

With maximum pedestrian density of about 0.26 ped/m² (80 people), the bridge is closest to Class III defined in Setra guideline. For this class the multiplication factor is

$$10.8\sqrt{\varsigma\cdot N}\cdot\Psi\cdot\left(\int_0^L|\phi(x)|dx\right)/L=10.8\cdot1\cdot\sqrt{0.0026\cdot N}\cdot39.0/104.0=0.21\sqrt{N}$$

as shown in Figure 8.12 (right, dashed line). Note that the factor is not to be used for small number of people (Class III footbridges are defined as those with around 0.5 ped/m² i.e. approximately 156 pedestrians on Podgorica bridge, which was never achieved during measurements).

For resonant stationary force with amplitude of 280N the maximum response is 0.93 m/s². Therefore, for 80 people on the bridge the estimated response would be $0.21\sqrt{80} \cdot 0.93 = 1.75$ m/s². However, if a moving single person model producing 0.45 m/s² peak acceleration is used, the predicted response would be 0.85 m/s², which is in excellent agreement with maximum measured level of 0.79 m/s².

If EC1 NA [21] is used instead, a multiplication factor of $0.35\sqrt{N}$ would be obtained, meaning that the NA results in conservative estimates of the response in case of the Podgorica footbridge.

8.5.3.2.3   Monte Carlo simulations

Monte Carlo simulations for 4.5 hours of pedestrian traffic have also been conducted to compare simulated with measured results. The arrival time from both sides was assumed to be 7 people per minute, a little above the experimentally observed total of 11 crossings per minute. Each pedestrian was modelled according to model defined in [16]. Distribution of amplitudes of main harmonics of the walking force as well as pacing frequency was assumed to be Gaussian. Taking into account modal properties of the relevant vibration mode, an overestimation of the response was obtained [38].

However, it seems that HSI occurs on this bridge when pedestrians feel perceptible vibrations i.e. they damp them out, so that even for a single person walking the damping is approximately doubled [57], hence new simulations of the bridge response with damping level increased from 0.26 to 0.6% were conducted. The result in the form of cumulative distribution for peak acceleration per cycle is shown in Figure 8.12 (left, dashed line). It can be seen that this distribution is very close to the measured one (solid line).

8.5.3.3   Changi Mezzanine Bridge

In this section the vertical response estimates for CMB, described previously in section 8.5.2.3, are presented.

8.5.3.3.1   Vertical response simulations based on single pedestrian models and synchronisation factor

Studies by the consultant [76], which were published before the final results from the LMB study were available and the SLE formula confirmed, used lateral forces of 25–50 N per pedestrian and vertical forces up to 280 N per pedestrian together with correlation coefficients multiplying the number of pedestrians to predict response levels. Various scenarios of crowd numbers between 200 and 1680 pedestrians on the bridge i.e. 0.25 to 2 pedestrians/m² produced estimates of response up to 3.3 m/s² for the lateral mode LS1 predicted (accurately) at 0.93 Hz and up to 11.5 m/s² for the torsional mode TS1 predicted at 1.52 Hz (and identified later at 1.64 Hz).

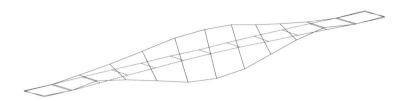

Figure 8.13 CMB Torsional mode at 1.64 Hz with 145 tonne modal mass and 0.4% damping.

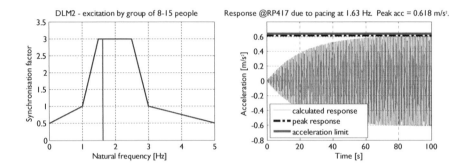

Figure 8.14 DLM2 prediction for CMB; at 0.62 m/s² six times the measured response levels for 150 pedestrians.

The study concluded that with more than 200 pedestrians the bridge would be perceived as lively, although the figure of 650 pedestrians was more representative of crowd loading. As a result pedestrian testing of the bridge was recommended and remedial measures suggested, as described previously.

In fact, testing of the bridge with three pedestrians prompted by metronome to walk at exactly the correct frequency generated vibration levels almost identical to those with 150 pedestrians, i.e. about 5% less than the BS5400 predictions of 0.1 m/s².

8.5.3.3.2 Vertical response simulations using Setra and proposed Eurocode models

The target for these simulations is the mode TS1 (Figure 8.13) which is lively for comfortable pacing rates.

The pedestrian density during the test with 150 people was 150/840 = 0.18 ped/m², which is well below the densities assumed for different classes of bridges in Setra guidance. However, if the procedure prescribed for response calculation for Class II/III footbridges is followed using the actual pedestrian number, then the estimate of peak response is around 0.31 m/s², with adjustment of the procedure to the fact that the mode shape is torsional rather than vertical. This is important as there is no explicit provision for torsional modes in the Setra guidelines, so some level of interpretation is required. This appears to be currently the most rational estimate of the vertical response using the models for crowds available, although it is still around three times larger than the peak measured response.

The prediction of DLM2 from the unadopted Eurocode 1 (i.e. FIB model [27]) proposal, which was supposed to represent the effect of a small group of 8–15 people, is shown in Figure 8.14. Once again, this is heavily conservative.

The DLM models (for Eurocode 1) also did not provide reliable results when applied to well known footbridges such as Solferino Bridge and so "were not proposed for the formal vote" [26].

8.5.3.3.3 Vertical response simulations using frequency domain approach

The previously presented time-domain approaches appear to be very sensitive to the assumptions made, in particular those which make use of multiplication factors, often yielding huge overestimation of the responses, as verified though some experimental response measurements.

To improve this, observation of the 'leakage' of DLF lines in Fourier spectra of walking forces, as shown in Figure 8.1, led to the development of a frequency domain model using random vibration theory as a basis to simulate response to a diffuse crowd of pedestrians [40].

The auto spectral density (ASD) of response $x(t)$ for a single mode is obtained using the mode frequency response function H(f) and the ASD of the pedestrian loading function.

$$S_X(f) = \psi_z^2 \left|H(f)\right|^2 S_{P,1}(f) \int_0^L \int_0^L \psi_{z_1} \psi_{z_2} coh(f, z_1, z_2) dz_1 dz_2 \tag{5}$$

where

$$S_{P,1}(f) = \left(NW/L\right)^2 \phi(f) \cdot G_1^2(f)/2 \tag{6}$$

is the ASD of the pedestrian force, obtained from a combination of $G_1^2(f)$, the distribution of spectral lines about the mean pacing frequency for a pedestrian and the probability density function $\phi(f)$ of pedestrian pacing rates. W is the weight of a pedestrian and L the length of the bridge used by pedestrians.

The mode shape is accounted for via ψ and a crucial part of the analysis is the coherence function $coh(f, z_1, z_2)$ between pedestrians at different locations. Integrating response ASD around the resonance leads to a dependence on $\sqrt{N/\zeta}$. This is the same as the result obtained empirically in the Setra study for Class II and III (Table 1), having in mind that Setra gives multiplication factors related to the force.

The resulting prediction of vertical responses is shown in Figure 8.15 which assumes zero coherence (even when SLE was observed). The predicted response is only slightly less than the observed counterpart, except for the period after 870 seconds when the 150 pedestrians began to walk after a stop, which suggests a certain degree of correlation.

The interesting observation that a model with zero correlation, even during observed 'SLE' provides a reasonable response estimate, clearly highlights the need

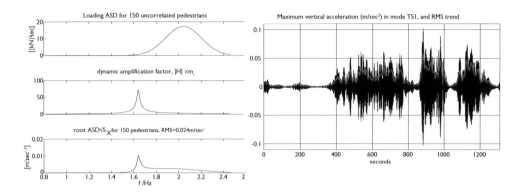

Figure 8.15 Frequency domain approach to predicting crowd-induced response (left) and correspondence with measured response at CMB (right).

for research on the nature of crowd synchrony and on the mechanism of SLE, which for initiation may not in fact need pedestrians to be synchronised!

A quick check on the pedestrian Scruton number (for vertical response) shows that for 150 pedestrians over 900 tonnes of bridge deck at 0.4% damping shows the value to be 0.6, three times higher than for LMB with the same number of pedestrians. Although this value is higher than the provisional limit of 0.27 [52] the bridge is lively and it seems that more research is necessary to establish a reliable limiting value for Scruton number.

8.6 CONCLUSIONS AND SUGGESTIONS

From this review of available literature, codes and standards related to crowd dynamic loading of footbridges and several experimental/analytical studies, there are several observations that point to future developments.

1   The most practically useful design guidance about vibration levels due to crowd loading is at present found in the Setra design guide. The model for calculating response due to dense crowds embodies a dependence on square root of both the number of pedestrians (N) and inverse of damping ratio (ζ). This is in line with a model that is based on random vibration theory and probability density functions of pedestrian loading. The Setra procedure provides the best match with observations and suggests further development of the frequency domain random vibration model.

2   For vibration response in the vertical direction on Podgorica footbridge and CMB, the BS5400-like 'bad man' prediction is adequately conservative even for a significant number of pedestrians, but under the condition that an appropriate DLF is accounted for. For example, the values suggested in [14] seem to be appropriate. On the other hand the 'bad man' approach does not work for the shorter

walkway, neither does it work for another short span footbridge with first mode susceptible to second harmonic forcing [69]. Relying on such an irrational code and hoping for its usual conservatism to apply is hardly due diligence in design.

3   While many theories have been proposed to explain synchronous lateral excitation (SLE), apparently none has been conclusively proven (in print, by December 2008). Meanwhile, the simple and semi-empirical Arup formula continues to agree well with observations on a significant number of bridges. Notions of some degree of pedestrian synchronisation being required to establish the condition (rather than to take over and grow response at some critical vibration level) are apparently sunk by lack of evidence of synchronisation on at least two bridges as reported by Macdonald [57].

4   There is abundant evidence of human-structure interaction (HSI) in the vertical direction which has at least two effects. HSI has the potential to increase vertical mode damping and to reduce pedestrian dynamic loads at large response levels (a form of lock-out rather than lock-in), which is, interestingly, opposite to what is happening in the horizontal lateral direction.

5   Data on pedestrian ground reaction forces (GRFs) are sparse, particularly for moving surfaces, and there has been no convincing study of variation of both vertical and lateral GRFs with structure motion.

6   Crude models of HSI are available, but these need further development to fit observations of reduced loading for high vertical response.

7   There is no known mechanism that accounts for pedestrian synchronisation in crowds. This is one area that urgently needs further research, though collaboration among engineering, psychology and biomechanics communities at least.

If the convenient random conservatism of the single 'bad man' approach is rejected, the only options currently viable for crowd loading dynamic simulations use a multiplication factor (for vertical response) or critical number of people (for lateral stability). Such approaches are convenient due to simplicity, but this oversimplification brings the danger of missing factors that could be of interest for a particular structure e.g. multi-mode response of a bridge, mode shapes not similar to sine functions, changeable damping, etc.

In these cases, at the present level of understanding it is recommended to use time domain Monte Carlo simulations which do not try to model over-simplistically a poorly understood process such as synchronisation among pedestrians.

The core of this approach should be a single pedestrian model capable of taking into account the complete frequency content of the walking force. Randomness in the load induced by crowd of people could be taken into account using distribution function for forcing amplitude and walking frequency, with uniform distribution of phase between different harmonic components.

Ultimately, effects of human-structure interaction and lock-in should be included, perhaps in the same way as proposed in the new UK guidance for crowd-induced response of grandstands [78].

Monte Carlo simulations are, however, expensive and beyond the ability of mainstream structural engineering consultants. An alternative approach that holds promise is the frequency domain random vibration approach that is mature in the wind engineering community and for which design guidance is available [79]. However, the drawback could be that the estimation of human response to vibration, normally linked with the peak or 1s RMS values, will become more difficult since these two quantities are normally not a direct output of frequency domain procedures, but need to be derived by a further statistical analysis, in the same way that statistical (e.g. peak) factors were researched by the wind engineering community 40 years ago.

Regarding synchronous lateral excitation (SLE), the best model currently available is still the Arup SLE formula which has been validated in the studies reported here (as well by the original London Millennium Bridge research).

Finally, no amount of prediction can guard against fate and our strongest recommendation is that any high-profile footbridge should be quietly crowd-tested and that provision should be made for retrofit in case of poor performance.

REFERENCES

[1] Wolmuth, B., Surtees, J., "Crowd-related failure of bridges", *Civil Engineering*, Vol. 156, No. 3, pp. 116–123, 2003.

[2] McRobie, A., Morgenthal, G., "Risk Management for Pedestrian-Induced Dynamic of Footbridges", *International Conference on the Design and Dynamic Behaviour of Footbridges*, Paris, 2002.

[3] Bachmann, H., Ammann, W., *Vibrations in Structures Induced by Man and Machines*, International Association of Bridge and Structural Engineering, 1987.

[4] Pavic, A., Reynolds, P., "Vibration serviceability of long-span concrete building floors. Part 2: Review of mathematical modelling approaches", *Shock and Vibration Digest*, Vol. 34, No. 4, pp. 279–297, 2002.

[5] BSI, *Steel, concrete and composite bridges. Part 2: Specification for loads*, BS 5400-2: 2006, British Standards Institution, 2006.

[6] The Highway Engineering Division, *Ontario highway bridge design code*, The Highway Engineering Division, Ontario, 1983.

[7] Wyatt, T. A., *Design guide on the vibration of floors*, The Steel Construction Institute, Construction Industry Research and Information Association, 1989.

[8] Allen, D.E., Murray, T.M., "Design criterion for vibrations due to walking", *Engineering Journal AISC*, Vol. 30, No. Part 4, pp. 117–129, 1993.

[9] Pavic, A., Willford, M.R., "Appendix G: Vibration serviceability of post-tensioned concrete floors", *Post-tensioned concrete floors design handbook*, pp. 99–107, Concrete Society, 2005.

[10] Smith, J.W., *Vibration of Structures*, Chapman and Hall, 1988.

[11] Ellis, B.R., "On the response of long-span floors to walking loads generated by individuals and crowds", *The Structural Engineer*, Vol. 10, No. 78, pp. 17–25, 2000.

[12] Bachmann, H., *Vibration Problems in Structures: Practical Guidelines*, Birkhäuser Verlag, 1995.

[13] Zivanovic, S., Pavic, A., Reynolds, P., "Vibration serviceability of footbridges under human-induced excitation: a literature review", *Journal of Sound and Vibration*, Vol. 279, No. 1–2, pp. 1–74, 2005.

[14] Kerr, S. C., *Human induced loading on staircases*, PhD Thesis. University College London, Mechanical Engineering Department, London, UK, 1998.

[15] Willford, M., Young, P., Field, C., "Predicting footfall-induced vibration: Part 1", *Structures & Buildings*, Vol. 160, No. SB2, pp. 65–72, 2007.

[16] Zivanovic, S., Pavic, A., Reynolds, P., "Probability-based prediction of multi-mode vibration response to walking excitation", *Engineering Structures*, Vol. 29, No. 6, pp. 942–954, 2007.

[17] Willford, M., "Dynamic Actions and Reactions of Pedestrians", *International Conference on the Design and Dynamic Behaviour of Footbridges*, Paris, 2002.

[18] Ricciardelli, F., Pizzimenti, D., "Lateral walking-induced forces on footbridges", *Journal of Bridge Engineering*, Vol. 12, No. 6, pp. 677–688, 2007.

[19] Ronnquist, A., *Pedestrian induced lateral vibrations of slender footbridges*, Norwegian University of Science and Technology, Trondheim, Norway, 2005.

[20] Setra, *Guide méthodologique passerelles piétonnes (Technical guide Footbridges: Assessment of vibrational behaviour of footbridges under pedestrian loading)*, Setra, 2006.

[21] BSI, *UK National Annex to Eurocode 1: Actions on structures - Part 2: Traffic loads on bridges*, NA to BS EN 1991-2: 2003, British Standards Institution, 2008.

[22] Matsumoto, Y., Nishioka, T., Shiojiri, H., Matsuzaki, K., "Dynamic design of footbridges", *IABSE Proceedings*, No. P-17/78, pp. 1–15, 1978.

[23] Wheeler, J.E., "Prediction and control of pedestrian induced vibration in footbridges", *Journal of the Structural Division ASCE*, Vol. 108, No. ST9, pp. 2045–2065, 1982.

[24] Fujino, Y., Pacheo, B.M., Nakamura, S.-I., Warnitchai, P., "Synchronisation of human walking observed during lateral vibration of a congested pedestrian bridge", *Earthquake Engineering and Structural Dynamics*, Vol. 22, No. 9, pp. 741–758, 1993.

[25] Dallard, P., Fitzpatrick, T., Flint, A., Low, A., Ridsdill Smith, R., Willford, M.R., Roche, M., "London Millenium Bridge: Pedestrian-Induced lateral Vibration", *ASCE Journal of Bridge Engineering*, Vol. 6, No. 6, pp. 412–417, 2001.

[26] Calgaro, J-A., "Designing footbridges with Eurocodes", *Eurocodesnews*, Vol. March, No. 2, pp. 6–6, 2004.

[27] FIB, *FIB bulletin 32: Guidelines for the design of footbridges.*, Federation Internationale du Beton, 2005.

[28] Barker, C., Mackenzie, D., "Design methodology for pedestrian induced footbridge vibrations", *3rd International Conference Footbridge*, Porto, 2008.

[29] Kerr, S. C., Bishop, N. W. M., "Human induced loading on flexible staircases", *Engineering Structures*, Vol. 23, pp. 37–45, 2001.

[30] Pachi, A., Ji, T., "Frequency and velocity of people walking", *The Structural Engineer*, Vol. 83, No. 3, pp. 36–40, 2005.

[31] Sahnaci, C., Kasperski, M., "Random loads induced by walking", *Sixth European Conference on Structural Dynamics (EURODYN)*, Paris, 2005.

[32] Zivanovic, S., Racic, V., El Bahnasy, I., Pavic, A., "Statistical characterisation of parameters defining human walking as observed on an indoor passerelle", *EVACES 2007*, Porto, Portugal, 2007.

[33] Ricciardelli, F., Briatico, C., Ingolfsson, E.T., Georgakis, C.T., "Experimental validation and calibration of pedestrian loading models for footbridges", *EVACES 2007*, Porto, Portugal, 2007.

[34] Occhiuzzi, A., Spizzuoco, M., Ricciardelli, F., "Loading models and response control of footbridges excited by running pedestrians", *Structural Control and Health Monitoring*, Vol. 15, No. 3, pp. 349–368, 2008.

[35] Kasperski, M, Sahnaci, C., "Serviceability of pedestrian structures xxposed to vibrations during marathon events", *26th International Modal Analysis Conference (IMAC XXVI)*, Orlando, 2008.

[36] Ingolfsson, E. T., Georgakis, C. T., Svendsen, M. N., "Vertical footbridge vibrations: details regarding experimental validation of the response spectrum methodology", *Footbridge 2008*, Porto, Portugal, 2008.

[37] Georgakis, C.T., Ingolfsson, E.T., "Vertical footbridge vibrations: the response spectrum methodology", *Footbridge 2008*, Porto, 2008.

[38] Zivanovic, S., Pavic, A., Reynolds, P., Vujovic, A., "Dynamic analysis of lively footbridge under everyday pedestrian traffic", *EURODYN 2005*, Paris, 2005.

[39] Simiu, E., Scanlan, R.H., *Wind effects on structures*, John Wiley and Sons,1996.

[40] Brownjohn, J.M. W., Pavic, A., Omenzetter, P., "A spectral density approach for modelling continuous vertical forces on pedestrian structures due to walking", *Canadian Journal of Civil Engineering*, Vol. 31, No. 1, pp. 65–77, 2004.

[41] Butz, C., "A probabilistic engineering load model for pedestrian streams", *Footbridge 2008*, Porto, 2008.

[42] McRobie, A., Morgenthal, G., Lasenby, J., Ringer, M., "Section model tests on human-structure lock-In", *Bridge Engineering*, Vol. 156, No. BE2, pp. 71–79, 2003.

[43] Hobbs, R. E., *Tests on lateral forces induced by pedestrians crossing a platform driven laterally*, Imperial College London, 2000.

[44] Pavic, A., Reynolds, P., Wright, J., *Analysis of frequency response functions measured on the Millennium Bridge*, University of Sheffield, 2001.

[45] Pavic, A., Willford, M., Reynolds, P., Wright, J. R., "Key results of modal testing of the illennium Bridge, London", *International Conference on the Design and Dynamic Behaviour of Footbridges*, 2002.

[46] Pavic, A., Armitage, T, Reynolds, P., Wright, J.R., "Methodology for modal testing of the Millennium Bridge, London", *Structures and Buildings*, Vol. 152, No. 2, pp. 111–121, 2002.

[47] Fitzpatrick, A., Dallard, P., le Bourva, S., Low, A., Ridsill Smith, R., Willford, M., *Linking London: The Millennium Bridge*, The Royal Academy of Engineering, 2001.

[48] Dallard, P., Fitzpatrick, A.J., Flint, A., le, Bourva S., Low, A., Ridsdill Smith , R., Willford, M., "The London Millennium Footbridge", *The Structural Engineer*, Vol. 79, No. 22, pp. 17–33, 2001.

[49] Piccardo, G., Tubino, F., "Parametric resonance of flexible footbridges under crowd-induced lateral excitation", *Journal of Sound and Vibration*, Vol. 311, No. 1–2, pp. 353–371, 2008.

[50] Roberts, T.M., "Lateral pedestrian excitation of footbridges", *ASCE Journal of Bridge Engineering*, Vol. 10, No. 1, pp. 107–112, 2005.

[51] Nakamura, S., "Model for lateral excitation of footbridges by synchronous walking", *Journal of Structural Engineering*, Vol. 130, No. 1, pp. 32–37, 2004.

[52] Newland, D.E., "Pedestrian excitation of bridges", *Journal of Mechanical Engineering Science*, Vol. 218, No. 5, pp. 477–492, 2004.

[53] Blekherman, A.N., "Autoparametric resonance in a pedestrian steel arch bridge: Solferino bridge, Paris", *ASCE Journal of Bridge Engineering*, Vol. 12, No. 6, pp. 669–676, 2007.

[54] Venuti, F., Bruno, L., Bellomo, N., "Crowd dynamics on a moving platform: Mathematical modelling and application to lively footbridges", *Mathematical and Computer Modelling*, Vol. 45, No. 3–4, pp. 252–269, 2007.

[55] Strogatz, S. H., Abrams, D. M., McRobie, A., Eckhardt, B., Ott, E., "Crowd synchrony on the Millennium Bridge", *Nature*, Vol. 438, pp. 43–44, 2005.

[56] Barker, C., "Some observations on the nature of the mechanism that drives the self-excited lateral response of footbridges", *International Conference on the Design and Dynamic Behaviour of Footbridges*, Paris, 2002.

[57] Macdonald JHG, "Lateral excitation of bridges by balancing pedestrians", Proceedings of the Royal Society A, First Cite, 2008.

[58] Zivanovic, S., Pavic, A., Reynolds, P., "Human–structure dynamic interaction in footbridges", *Proceedings of ICE, Bridge Engineering*, Vol. 158, No. BE4, pp. 165–177, 2005.

[59] Brownjohn, J.M.W., Fok, P., Roche, M., Moyo, P., "Long Span Steel Pedestrian Bridge at Singapore Changi Airport - part 1: Prediction of Vibration Serviceability Problems", *The Structural Engineer*, Vol. 82, No. 16, pp. 21–27, 2004.

[60] Yao, S., Wright, J.R., Pavic, A., Reynolds, P., "Experimental study of human-induced dynamic forces due to jumping on a perceptibly moving structure", *Journal of Sound and Vibration*, Vol. 296, No. 1–2, pp. 150–165, 2006.

[61] Zivanovic, S., Diaz, I.M., Pavic A, "Influence of walking and standing crowds on structural dynamoc performance", *27th International Modal Analysis Conference (IMACXXVII)*, 2009.

[62] Newland, D.E., "Pedestrian excitation of bridges – recent results", *10th International Congress on Sound and Vibration*, Stockholm, Sweden, 2003.

[63] Macdonald, J. H. G., "Pedestrian-induced vibrations of the Clifton Suspension Bridge", *Proceeding of the Institution of Civil Engineers, Structures and Buildings*, Vol. 161, No. 2, pp. 69–77, 2008.

[64] Dziuba, P., Grillaud, G., Flamand, O., Sanquier, S., Tetard, Y., "La passerelle Solferino: Comportement dynamique (Solferino bridge: Dynamic behaviour)", *Bulletin ouvrages metalliques*, No. 1, pp. 34–57, 2001.

[65] Brownjohn, J.M.W., Fok, P., Roche, M., Omenzetter, P., "Long Span Steel Pedestrian Bridge at Singapore Changi Airport – part 2: Crowd Loading Tests and Vibration Mitigation Measures", *The Structural Engineer*, Vol. 82, No. 16, pp. 28–34, 2004.

[66] Caetano, E., Cunha, A., Moutinho, C., "Implementation of passive devices for vibration control at Coimbra footbridge", *EVAES07*, Porto, 2007.

[67] Nakamura, S., Kawasaki, T., "Lateral vibration of footbridges by synchronous walking", *Journal of Constructional Steel Research*, Vol. 62, pp. 1148–1160, 2006.

[68] Zivanovic, S., Brownjohn, J.M.W., Pavic, A., "Vibration performance of footbridges under pedestrian traffic", *Third International Conference Footbridge 2008*, Porto, 2008.

[69] Brownjohn, J.M.W., Middleton, C., "Efficient dynamic performance assessment of a footbridge", *Proceedings of ICE, Bridge Engineering*, Vol. 158, No. BE4, pp. 185–192, 2005.

[70] Brownjohn, J.M.W., Tao, N.F., "Vibration excitation and control of a pedestrian walkway by individuals and crowds", *Journal of Shock and Vibration*, Vol. 12, No. 5, pp. 333–347, 2005.

[71] Song, S.K., Zhao, L., Wong, K.S., "Dynamic analysis and vibration measurements of the Tanjong Rhu Suspension Footbridge, Singapore", *EVACES 05*, Bordeaux, France, 2005.

[72] Sudjic, D., *Blade of light: The story of London's Millennium Bridge*, Penguin press, 2001.

[73] Dallard, P., Fitzpatrick, A., Flint, A., Low, A., Ridsill Smith, R., "The Millennium Bridge, London: problems and solutions", *The Structural Engineer*, Vol. 79, No. 8, pp. 15–17, 2001.

[74] Dougill, J.W., Wright, J.R., Parkhouse, J.G., Harrison, R.E., "Human structure interaction during rhythmic bobbing", *The Structural Engineer*, Vol. November 2006, No. 22, pp. 32–39, 2006.

[75] Zivanovic, S., Pavic, A., Brownjohn, J.M.W., "Vibration serviceability assessment of footbridges using VSATs software", *EURODYN 2008*, Southampton, UK, 2008.

[76] Zivanovic, S., Pavic, A., "Probabilistic approach to subjective assessment of footbridge vibration", *The 42rd UK Conference on Human Responses to Vibration*, Southampton, UK, 2007.

[77] Roche, M., *Land Transport Authority Singapore Changi Bridge vibration characteristics*, 2000.

[78] Pavic, A., Reynolds, P., "Experimental verification of novel 3DOF model for grandstand crowd-structure interaction", *26th International Modal Analysis Conference (IMAC XXVI)*, 2008.

[79] ESDU, *Calculation methods for along wind loading Part 3. Response of buildings and plate-like structures to atmospheric turbulence.*, IHS ESDU International, 2007.

Chapter 9

Application of tuned mass dampers for bridge decks

Christian Meinhardt
Gerb Vibration Control Systems, Essen, Germany

SUMMARY

This contribution gives an introduction about the design and the practical application of Tuned Mass Dampers (TMD). Especially the practical adaptation of the theoretical defined optimum specification and the effect of all relevant parameters will be discussed. On the basis of realized footbridge projects where TMDs have been successfully applied to reduce the occurring vibrations, practical ways for an experimental determination of the relevant dynamic behaviour and the in situ assessment of the TMD effectiveness will be introduced.

Keywords: Footbridge, Damping Ratio, Tuned Mass Damper; TMD Specification.

9.1 INTRODUCTION – BASICS ON TMDS

Due to the trend of constructing ever lighter and filigree load-carrying structures, footbridges are becoming more susceptible to vibrations caused by pedestrians or wind. Usually, these vibrations impact only the serviceability of the bridges since the desired level of comfort is no longer attained. However, in some cases, the vibrations of the bridges are so extreme that damages can arise or, in extreme cases, the structural integrity of the bridge can be at risk.

The primary reason for the occurrence of perturbing vibrations is resonance. That means the vibrations only happen for an excitation of the structure with a frequency which is similar to the natural frequency or a multiple of it. A complex structure exhibits more than one natural frequency. The number of natural frequencies that are in range of those which will be excited by wind- or human- induced vibrations depends on the engineering design, the dimensions and material of the structure.

Each natural frequency is associated with a defined natural mode which characterizes the swinging behaviour of the structure. Natural modes can be bending modes in vertical and horizontal direction or torsional modes. They appear in different orders, where the order of the mode exposes the number of minima and maxima for the certain kind of bending or torsional mode. The higher the order of the natural mode, the higher is the associated natural frequency.

For the case that an excitation causes a resonance-like vibration state, the damping associated with the natural mode of the considered structure is relatively small. Damping is a combination of material damping, depending on the used material, and on the structural damping which is defined by the structure's constructive design. Concrete structures reveal a higher damping than steel structures, not only because concrete exhibits a higher material damping, but also because the interaction between concrete and reinforcement increases the general damping. Welded steel structures, for example, are supposed to exhibit the lowest damping. The small damping in the case of a resonance-like excitation results in a slow attenuation of the vibrations.

In order to improve the dynamic behaviour of a footbridge, the structural stiffness of the bridge has to be increased so that its natural frequencies are out of the range

Figure 9.1 Applied TMD at a footbridge.

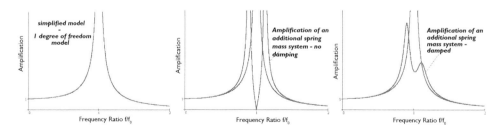

Figure 9.2 Effect of a Tuned Mass Damper.

typically excited by pedestrians or wind, or else structural damping of the bridge has to be increased. The most effective way to increase the structural damping and improve the dynamic behaviour is the application of dampers (buffers, shock absorbers, Dampers) or TMDs (Tuned Mass Dampers) that significantly reduce resonance-like excitation related vibrations (see Fig. 9.1).

A TMD is a vibrating mass that displays movements which are contrary to those of the main structure. To attain these contrary movements, the mass is elastically supported and tuned for the frequency that has to be eliminated. The contrary movements cause inertial forces that compensate the structures movements by depriving vibration-energy from the structure. That leads to an increase of damping. Additionally the interaction between TMD and structure causes a subdivision of the natural frequencies (Fig. 9.2).

The advantage of using TMDs is that, unlike with the use of dampers, shock absorbers etc., no fixed-point are required. It is simply force-fitted to the structure. Compared to the additional structural mass that is required to increase the structural damping conventionally, the necessary TMD mass is only a small fraction, which allows also a subsequent application with no bigger improvements in the structural design.

The following will give an introduction about an evaluation of footbridges regarding their vibration susceptibility especially to human induced vibrations and will define empirical values to estimate structural damping ratio of footbridges.

For an application of Tuned Mass Dampers the definition of the optimum specification is discussed and compared to practical conditions. The design of TMDs will also be introduced. Finally some example projects will be presented.

9.2   PRACTICAL ADAPTIVE TMD SPECIFICATIONS – OPTIMIZATION

The TMD specifications – effective mass, tuning frequency and TMD damping ratio-can be determined regarding a 2 degree of freedom model (see Fig. 9.3).

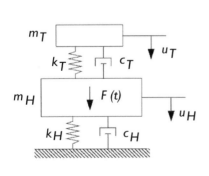
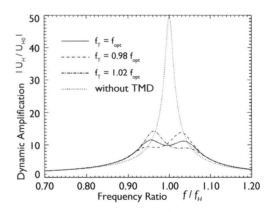

Figure 9.3 Two degree of freedom model – amplification functions for several tuning frequencies.

$$m_H \ddot{u}_H + c_H \dot{u}_H + c_T(\dot{u}_H - \dot{u}_T) + k_H u_H + k_T(u_H - u_T) = F(t)$$
$$m_T \ddot{u}_T + c_T(\dot{u}_T - \dot{u}_H) + k_T(u_T - u_H) = 0 \tag{1}$$

An amplification function can be derived by using the equations of motion for the coupled structural mass and the TMD mass (1), by applying an exponential approach (2), (3) and simplifying the equation using the terms that are displayed in (4).

$$u_T = U_T e^{i\omega t} \qquad F(t) = F_H e^{i\omega t} \tag{2}$$

$$\left[-\omega^2 m_H + i\omega(c_H + c_T) + (k_H + k_T) \right] U_H + \left[-i\omega c_T - k_T \right] U_T = F_H$$
$$\left[-i\omega c_T - k_T \right] U_H + \left[-\omega^2 m_T + i\omega c_T + k_T \right] U_T = 0 \tag{3}$$

$$\omega^2 \frac{m_H}{k_H} = \frac{\omega^2}{w_H^2} = \Omega^2$$

$$\omega \frac{c_H}{k_H} = \omega \frac{2\zeta_H \omega_H m_H}{k_H} = 2\zeta_H \frac{\omega}{\omega_H} = 2\zeta_H \Omega$$

$$\omega \frac{c_T}{k_H} = \omega \frac{2\zeta_T \omega_T m_T}{k_H} \frac{m_H}{m_H} = 2\zeta_T \frac{\omega_T}{\omega_H} \frac{\omega}{\omega_H} \frac{m_T}{m_H} = 2\zeta_T \beta\gamma\Omega \tag{4}$$

$$\frac{k_T}{k_H} = \frac{\omega_T^2}{\omega_H^2} \frac{m_T}{m_H} = \beta^2\gamma$$

$$\omega^2 \frac{m_T}{k_H} = \omega^2 \frac{m_T}{m_H} \frac{m_H}{k_H} = \frac{\omega^2}{\omega_H^2} \frac{m_T}{m_H} = \Omega^2\gamma$$

$$\frac{F_H}{k_H} = U_{H0}$$

Introducing dimensionless terms (5) and identifying the natural frequencies ω_H, ω_T using (6) leads to the system of equations (7) that can be used to calculate the amplification functions for the deflection U_{H0} under a static load F_H displayed in Fig. 9.5 for several tuning frequencies.

$$\beta = \frac{\omega_T}{\omega_H} = \frac{f_T}{f_H} \qquad \gamma = \frac{m_T}{m_H} \qquad \zeta_H \qquad \zeta_T \qquad \Omega = \frac{\omega}{\omega_H} \tag{5}$$

$$\omega_H = \sqrt{k_H / m_H} \qquad \omega_T = \sqrt{k_T / m_T} \tag{6}$$

$$\left[-\Omega^2 + 2i\Omega\left(\zeta_H + \beta\gamma\zeta_T\right) + \left(1 + \beta^2\gamma\right) \right] U_H + \left[-2i\Omega\beta\gamma\zeta_T - \beta^2\gamma \right] U_T = U_{H0}$$
$$\left[-2i\Omega\beta\gamma\zeta_T - \beta^2\gamma \right] U_H + \left[-\Omega^2\gamma + 2i\Omega\beta\gamma\zeta_T + \beta^2\gamma \right] U_T = 0 \tag{7}$$

Besides the tuning frequency f_T of the TMD and its Damping ratio c_T the TMD effect significantly depends on the ratio between the structures mass and the TMD-mass γ. By comparing the results, optimum values (minimum amplification) for the TMD specifications can be defined that depend from each other. An analytical optimization of the TMD specification considering all parameters becomes difficult. Therefore DEN HARTOG has specified a solution, disregarding the structural damping (8).

$$f_{opt} = \frac{f_H}{1 + m_T / m_H} \qquad \zeta_{opt} = \sqrt{\frac{3 m_T / m_H}{8\left(1 + m_T / m_H\right)^3}} \tag{8}$$

The diagram, shown in Figure 9.4, shows the curves for the optimized TMD parameters which should be only applied for harmonic excitations of the structure.

Regarding the practical application of TMDs, two problems have to be considered that influence the design of the dampers. On one hand it is difficult to achieve the optimized damping ratio for all conditions and for the life cycle of the structure that has to be protected. On the other hand, the dynamic loads, caused by the relative movements of the TMD shall be limited to minimize the impact to the structure as well as to guarantee the TMD performance for its life cycle. These supplementary conditions to the optimum TMD specification lead to the question how much the reduction effect of a TMD depends on the damping ratio and the tuning frequency, which also varies due to the nonlinearities, temperature or additional masses on the bridge.

Figure 9.5 shows the amplification functions for four variations of TMD specifications. The movements of the structure without and with TMD are shown as well as the movements of the TMD itself. The diagrams show that the TMD movements reduce with a bigger TMD mass. Also it can be seen, that the reduction effect increases with an increase of the TMD mass while the increase of the TMDs damping ratio does not effect the reduction due to the TMD.

The variation with a bigger TMD mass is also less affected by the shifted frequency. These trends can be summarized by the graphs that are displayed in Figure 9.6.

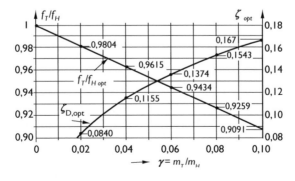

Figure 9.4 Optimization of TMD parameters [6].

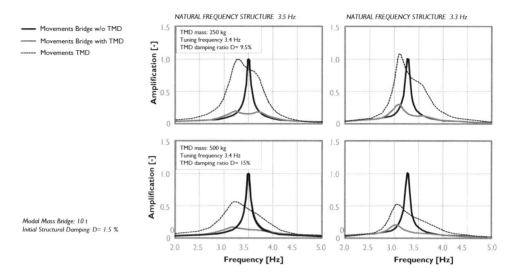

Figure 9.5 Variations of TMD specifications – resulting amplification functions.

The diagrams – withdrawn from an analytical two degree of freedom model - show that the Amplitude reduction, which is displayed by the factor that is characterized by the structures movement U_H and the movement of the bridge without a TMD U_0 strongly depends on the applied TMD mass. That also applies for the relative TMD movements U_T/U_0 which become smaller by using a higher TMD mass. The diagram also shows that a reasonable increase of the TMD damping ratio (<30%) compared to the optimum damping ratio ζ_{opt} does not effect the reduction U_H/U_0 while a undervalued damping ratio decreases the reduction effect and leads to bigger TMD movements U_T/U_0. So for a practical adaptation, the TMD ratio should be set greater that the optimum damping ratio to limit the TMD movements and to guarantee the effectiveness of the damping element (viscous damper) which might be subjected to fluctuations (temperature/humidity etc.).

Figure 9.6 also shows that the reduction effect is less influenced by a detuning – for example due to incorrect dynamic calculations that were used for the design or due to nonlinear ascendancies such as temperature or additional masses – when the tuning frequency is lower than the structures natural frequency. That means that a tuning frequency below the optimum frequency f_{opt} should be specified. Again the lower the tuning frequency regarding the structures natural frequency, the smaller are the relative movements.

The results of this theoretical examination have been considered for the design of TMDs in several projects. Particular aspects of the design components are described in the following chapter. Figure 9.7 shows the experimentally determined reduction ratio U_H/U_0 depending on the applied TMD mass related to the modal mass of the structure for conducted projects and displays a fair agreement with the theoretical results.

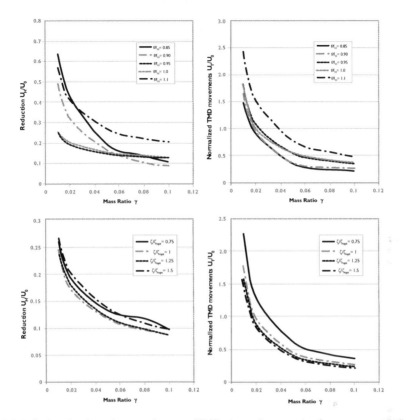

Figure 9.6 Left: Amplitude reduction due to a TMD depending on the frequency ratio f/f_H and the damping ratio ζ / ζ_{opt} – Right: Normalized TMD movements U_T/U_0 depending on the frequency ratio f/f_H and the damping ratio ζ / ζ_{opt}.

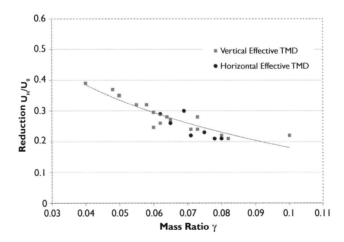

Figure 9.7 Experimentaly determined reduction ratio against the applied mass ratio for several conducted projects.

9.3   PRACTICAL ASSESSMENT OF THE DYNAMIC BEHAVIOUR – CONCEPTS FOR A TMD APPLICATION

For a TMD specification to successfully reduce human or wind induced vibrations, the following information are required:

- Natural frequency in which the footbridge is susceptible
- Corresponding mode shape (maximum of deflections)
- Corresponding modal mass
- Movements of the structure without TMD.

The uncertain parts this relevant information can be defined by dynamic calculations or more accurate, by in-situ measurements of the structures dynamic behaviour. Usually the natural frequencies, the corresponding modes and damping ratios are determined by an experimental modal analysis, where the modal displacements (displacements according to the frequency range) of a representing grid of measuring points are measured for an ambient excitation (see Fig. 9.8). Knowing the displacements and their phase for each characteristic frequency, the mode shapes can be displayed according to the geometric grid. This method is elaborate and needs a certain expenditure of time.

According to their construction type, footbridges display relevant mode shapes (vertical, horizontal and torsional) that can be generalized (see Fig. 9.9). To determine these expected generalized mode shapes, a more simplified testing method can be applied. To identify the relevant vertical mode shapes it is sufficient to measure the occurring vibrations due to an ambient or impulse-like excitation simultaneously at only 2–3 points in mid-span, quarter-span respectively third-span. Analyzing for which frequency the measuring points show bigger vibrations (see Fig. 9.10b), leads to an identification of mode shapes. The determination of the horizontal mode shapes happens analogously. Considering the phasing of the recorded time histories, the torsion modes can be identified as well (see Fig. 9.10a).

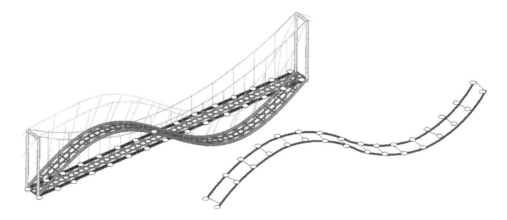

Figure 9.8 Required grid of measuring points for an experimental modal analysis.

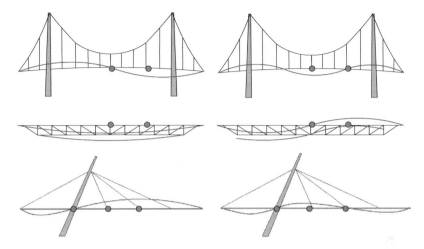

Figure 9.9 First two vertical deck mode shapes of several footbridge types – measurement point layout to capture these mode shapes.

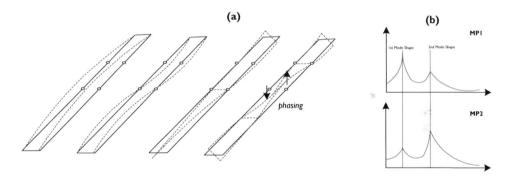

Figure 9.10 (a) First two horizontal an torsional deck mode shapes – measurement point layout to capture these mode shapes; (b) Frequency spectrum on two measuring points to identify the mode shapes.

Knowing the natural frequencies and the corresponding mode shapes the vibration susceptibility of the bridge deck can be assessed, exciting the bridge deck by jumping with a given beat according to the determined natural frequencies.

For each susceptible mode, the damping ratio can be determined analyzing the decaying behaviour of the bridge's vibrations.

In many cases it is required to apply TMDs for more than just one natural mode. Therefore the TMDs have to be placed at the locations of the biggest deflections for each mode (see Fig. 9.11). In case TMDs for different mode shapes are placed at the same location (vertical and torsional modes), the reduction effect due to the interaction of the TMDs can be estimated with the diagrams in Figure 9.6 subject to the tuning frequency difference of the TMDs and the modal masses for each mode shape.

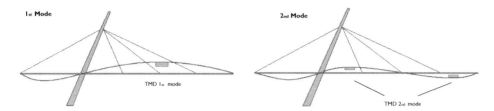

Figure 9.11 TMD layout for different vibration susceptible mode shapes.

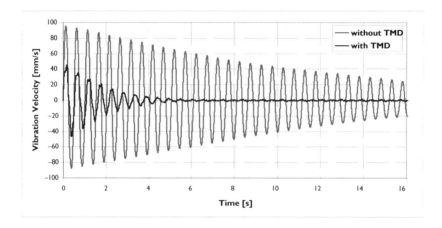

Figure 9.12 Time history of recorded vibrations for a defined load case without and with TMD.

To estimate the interaction effect for TMDs in different locations, more complex calculations that capture the precise mass distribution such as Finite Element Calculations are required.

The effectiveness of an applied TMD system can be determined with the same measurement campaigns that were used to determine the dynamic behaviour of the bridge deck. Therefore it is sufficient to assess the increase of structural damping due to the TMDs and to identify the occurring vibrations for a defined load case with and without TMD. Figure 9.12 shows the time history of recorded vibrations for a load case. It can be clearly seen that the vibrations reduced and the damping ratio increased due to the application of TMDs.

9.4 CONCLUSIONS

Due to the application of Tuned Mass Dampers the dynamic behaviour of footbridges can be enhanced to reduce the vibration susceptibility to human and wind induced vibrations. It has been shown, that the optimum TMD specifications have to follow practical aspects such as allowable TMD movements and the allowance of detuned

TMDs and variations of the damping ratio. This paper also documents the adequate results to assess the dynamic behaviour and the vibration susceptibility of footbridges with simplified measuring methods.

REFERENCES

[1] C. Petersen ,'Dynamik der Baukonstruktionen', *Vieweg Verlag*, 1996.
[2] J.P. Den Hartog,'Mechanische Schwingungen', 2. Edition, *Springer Verlag*, 1952.
[3] B. Weber, Lecture 'Tragwerksdynamik", *ETH Zürich*, 2002.
[4] H. Bachmann, W. Amman,'Schwingungsprobleme bei Bauwerken: Durch Menschen und Maschinen induzierte Schwingungen', *IABSE*, 1987.
[5] H. Bachmann, 'Vibration Problems in Structures- Practical Guidelines', *Birkhäuser Verlag*, 1995.
[6] C. Petersen, 'Schwingungsdämpfer im Ingenieurbau', *Maurer und Söhne GmbH*, 2001.
[7] H. Grundmann et al., 'Schwingungsuntersuchungen von Fußgängerbrücken', *Bauingenieur Vol. 68, Springer Verlag*, 1993.
[8] Y. Matsumoto et al., 'Dynamic Design of footbridges', *IABSE*, 1978.

Chapter 10

Experience and practical considerations in the design of viscous dampers

Philippe Duflot
Taylor Devices Europe, Brussels, Belgium

Doug Taylor
Taylor Devices Inc, North Tonawanda, USA

ABSTRACT

Modern pedestrian bridges tend to be long and slender in form, usually leading to a structural design with relatively low frequency primary modes of vibration. This type of structure can be excited to a nearly resonant response by various types of synchronized crowd activities and added damping is often required to prevent excessive structural motions and loadings.

This presentation provides a brief overview of fluid damping technology with specific case studies being provided from pedestrian bridges now equipped with fluid viscous dampers. The viscous dampers are used to completely suppress the feedback between the pedestrians and the bridge and/or wind-induced vibrations.

Design requirements of viscous damping devices for footbridge are discussed.

The tested performance of structures with fluid viscous dampers show that tremendous gains in performance can be realized at relatively low cost.

10.1 INTRODUCTION TO VISCOUS DAMPERS: DEFINITION AND FUNCTIONAL OUTPUT

A damper can be globally defined as an element which can be added to a system to provide forces which are resistive to motion, thus providing a means of energy dissipation.

The most convenient and common functional output equation for a damper can be characterized as:

$$F = C \cdot V^{\alpha}$$

Where F is the output force, V the relative velocity across the damper, C is the damping coefficient and α is a constant exponent which is usually a value between 0.3 and 2.

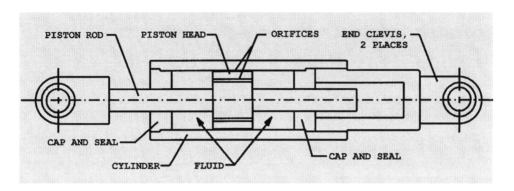

Figure 10.1 Viscous damper.

Fluid viscous dampers operate on the principle of fluid flow through orifices. A stainless steel piston travels through chambers that are filled with silicone oil. The silicone oil is inert, non flammable, non toxic and stable for extremely long periods of time. The pressure difference between the two chambers cause silicone oil to flow through orifices in the piston head and input energy is transformed into heat, which dissipates into the atmosphere.

Fluid viscous dampers can operate over temperature fluctuations ranging from −40°C to +70°C.

Notice that there is no spring force in this equation. Damper force varies only with velocity. For a given velocity the force will be the same at any point in the stroke. As dampers provide no restoring force the structure itself must resist all static loads.

The damper decreases the response by adding energy dissipation to a structure [1], which significantly reduces response to any vibration or shock inputs.

10.1.1 The effect of different values of α, the velocity exponent

Figure 10.2 shows the hysteresis loop of a pure linear viscous damper when subjected to a sinusoidal input. The loop is a perfect ellipse. The absence of storage stiffness makes the natural frequency of a structure incorporated with the damper remain the same. This advantage will simplify the design procedure for a structure with supplemental viscous dampers.

Fluid viscous dampers have the unique ability to simultaneously reduce both stress and deflection within a structure subjected to a transient. This is because a fluid viscous damper varies its force only with velocity, which provides a response that is inherently out-of-phase with stresses due to flexing of the structure.

Fig 10.3 shows a plot of force against velocity for several values of α, the velocity exponent. A value of $\alpha = 1$ is the curve for linear damping which is a good place to start in the design of a damping system. The hysteresis loop for a linear damper is a pure ellipse as shown in fig 10.2. $\alpha = 0.3$ is the lowest damping exponent normally

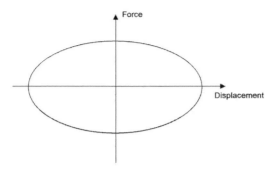

Figure 10.2 Hysteresis loop of viscous damper.

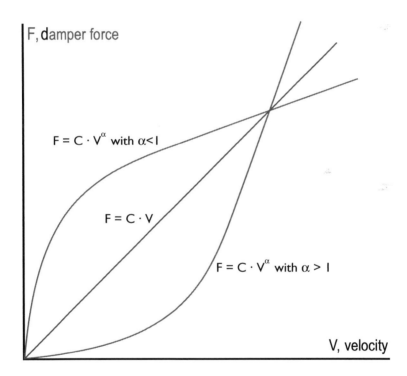

Figure 10.3 Force against velocity for different exponent values.

possible. Fig 10.3 shows this value provides significantly more force at lower velocities than a linear damper.

Linear damping is easy to analyse and can be handled by most software packages. Also linear damping is unlikely to excite higher modes in a structure.

Another advantage of linear damping is that there is very little interaction between damping forces and structural forces in a structure.

For most pedestrian bridges, linear dampers are therefore preferably considered to completely eliminate the biodynamic feedback between pedestrians and the bridge.

10.2   FLUID VISCOUS DAMPERS: DESIGN ELEMENTS

The essential design elements of a fluid damper are relatively few. However, the detailing of these elements varies greatly and can, in some cases, become both difficult and complex. Fig. 10.1 depicts a typical fluid damper and its parts. It can be seen that simply by moving the piston rod back and forth, fluid is forced through the piston head orifices, generating damping force.

Major part descriptions are as follows, using Fig. 10.1 as reference:

Piston rod: highly polished on its outside diameter, the piston rod slides through the seal and seal retainer. The external end of the piston rod is affixed to one of the two mounting clevises. The internal end of the piston rod attaches to the piston head. In general, the piston rod must react all damping forces, plus provide a sealing interface with the seal. Stainless steel is preferred as a piston rod material since any type of rust or corrosion on the rod surface can cause catastrophic seal failure. In some cases, the stainless steel must be chrome plated for compatibility with the seal material. For applications requiring a long stroke, a structural steel tube guide sleeve is used to protect the piston rod from excessive bending loads.

Cylinder: the damper cylinder contains the fluid medium and must accept pressure vessel loading when the damper is operating. Cylinders are usually manufactured from seamless steel tubing. Welded or cast construction is not permissible for damper cylinders, due to concerns about fatigue life and stress cracking.

Cylinders normally are designed for a minimum proof pressure loading equal to 1.5 times the internal pressure expected under a maximum credible dynamic input. By definition, the proof pressure must be accommodated by the cylinder without yielding, damage or leakage of any type.

Fluid: Dampers used in structural engineering applications require a fluid that is fire-resistant, non-toxic, thermally stable and which will not degrade with age. Typical silicone fluids have a flashpoint in excess of 340°C, are cosmetically inert, completely non-toxic and are among the most thermally stable fluids known to man. Since silicone fluids are produced by distillation, the fluid is completely uniform and no long-term settling will occur. Occasionally, other types of fluids are considered for special applications when necessary.

Seal: The seal materials must be carefully chosen for the service life requirement and for compatibility with the damper's fluid. Since dampers in footbridges can often be subject to long periods of infrequent use, seals must not exhibit long-term sticking nor allow slow seepage of fluid. Most dampers use dynamic seals at the piston rod interface and static seals where the end caps or seal retainers are attached to the cylinder. Dynamic seals for the piston rod should be manufactured from high-strength structural polymers to eliminate sticking or compression set during long periods of inactivity. Typical dynamic seals materials include TeflonR, stabilized nylon and members of the acetyl resin family. Dynamic seals manufactured from structural polymers do not age, degrade or cold flow over time.

Piston head: the piston head attaches to the piston rod and effectively divides the cylinder into two pressure chambers. As such, the piston head serves to sweep fluid through orifices located inside it, thus generating damping pressure.

Orifices: the pressurized flow of the fluid across the piston head is controlled by orifices. These can consist of a complex modular machined passageway or alternatively, can use drilled holes, spring loaded balls, poppets or spools. The use of any type of spring loaded orifice raises the reliability issues and proper performance.

Dependent on the shape and area of these passages, damping exponent ranging from 0.3 to 2.0 can be obtained without requiring any moving parts in the orifice.

10.3   PERFORMANCE OF VISCOUS DAMPERS

A properly designed viscous damper will attenuate transient and steady state inputs while staying within specified performance bounds. At the same time, the damper, as it is manufactured, must not yield, leak or overheat during use.

10.3.1   Transient inputs: frequency and response time

Depending on its end use, dampers can easily be designed and constructed to attenuate input transients in the range of 0–2000 Hz. However, for footbridge applications, spectral inputs rarely contain much content at frequencies in excess of 10 Hz. Using conventional mechanical engineering practice for vibrating systems, a control device should be capable of operating at a frequency of at least 10 times the maximum input frequency. Thus, a frequency response range of 0–100 Hz is sufficient for most footbridge damper applications.

Impulse and frequency response of full scale dampers built for a specific project can readily be determined by drop testing where a weight is allowed to free-fall a certain distance and then impact the damper.

10.3.2   Transient inputs: magnitude of required damping

The magnitude of viscous damping added to a structure for the suppression of vibration, wind or other transient inputs is usually in the range of 5–45% of critical. This is a very wide range and varies with the type of structure and excitation. Obviously the amount of damping selected is the responsibility of the engineer of record but generalized damping levels from previous projects are 15 to 25% added damping for footbridges subject to vibration and/or wind inputs [2].

10.3.3   Steady state inputs: wind and vibration

Footbridges normally use fluid dampers for the control or reduction of measured or felt vibration and wind responses.

As noted previously, viscous dampers are built to mitigate the response of inputs in the 0–100 Hz range. With respect to low amplitude vibrations, fluid dampers have been used to suppress amplitudes as low as 0.025 mm.

10.3.4   Heating effects

The thermal response inside a damper must be calculated to prevent overheating of internal parts during use. In most cases, overheating damage manifests itself by leakage, usually caused by a softened or melted dynamic seal.

Heat transfer calculations for footbridge dampers are relatively complex and the damper manufacturer must be given generalized motion data to properly size the damper. In general, manufacturers allow steady state heating of the damper to be no more than 40°C over ambient. If calculations indicate that overheating is an issue, then in most cases the damper will be increased in physical envelope until temperature rise during operation is low enough so as to be safely accepted by the internal parts.

10.3.5   Cyclic Life – Service Life

A properly designed and manufactured damper should not require any type of periodic service. The warranty should state that no periodic fluid replenishment or periodic servicing of any type is required.

The type of dynamic seal used in dampers is limited in life by wearing of the seal as the piston rod moves back and forth. In general, seal life is measured in terms of the total number of meters of rod displacement during a damper's lifetime.

10.4   CUSTOMER CONTROLLED PARAMETERS

Fluid dampers for a specific project are essentially adjusted by the manufacturer to meet specific customer specified parameters. The parameters include:

1   maximum rated force
2   minimum safety factors to yield
3   minimum required usable deflection from neutral position
4   damping constant
5   damping exponent
6   operating temperatures
7   maximum power input
8   maximum damper envelope
9   damping mounting configuration.

The maximum rated force of the damper is usually the force expected during the maximum credible event that the device is designed for. The safety factor to yield is based upon either the maximum rated force or the velocity at which this maximum force occurs. Typically, the safety factor is 1, 5 to 1, meaning that the damper will not yield when subjected to a force or velocity 150% of the rated maximum.

The minimum required usable stroke or deflection is the minimum distance that the damper should be able to stroke from its neutral position taking into account the dynamic stroke (due to dynamic inputs: vibration, wind, accidental input, ...) and the static stroke (due to thermal expansion/contraction, potential misalignment of the damper due to various tolerances of the bridge construction, ...)

The damping constant, damping exponent and temperature ranges can be easily expressed on a graph, defining allowable damper performance bandwidth at any defined operating temperature. Fig 10.4 provides a damper performance graph for an application on a footbridge.

The damper is tested in compression and in tension at different velocities as expected by the application. Force and displacement are recorded, the velocity is read at the peak force to eliminate the rise time effect. The force-velocity graph is then generated. The function must be within the foreseen tolerances, usually +/−15%.

Note that no significant airlag, backlash or compressibility is acceptable.

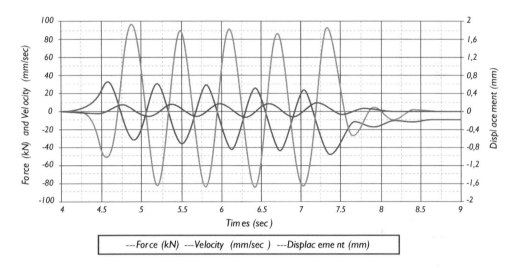

Figure 10.4 Example of force, velocity and displacement vs time measured on a footbridge damper.

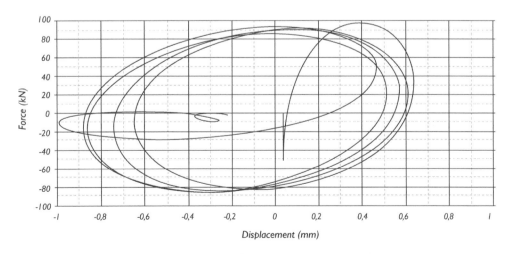

Figure 10.5 Force vs displacement.

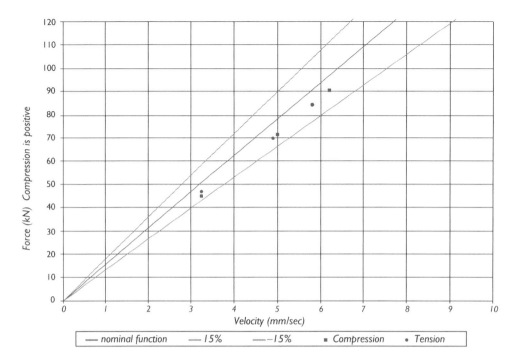

Figure 10.6 Force vs velocity.

A nominal function is the middle line in the plot with high and low tolerance limits applied. Operating temperature is 0°C to 60°C and the damper's output must fall within the +/–15% tolerance at all velocities, at any defined operating ambient temperature, both before and after the damper has absorbed the energy of a maximum credible event.

Performance at small amplitude of the damper must be controlled. The static friction is usually limited to 1% of total damper force.

Specific applications may impose diameter or length restrictions on the damper and these maximum values can also be customer specified. In most cases, a diameter limitation is much more common than a length restriction.

The usual way to attach dampers to a bridge is to bolt them to the structure. The dampers are equipped with maintenance free spherical bearings at each end and are connected to the bridge using mounting pins. The typical plus or minus 5° rotation angle of a spherical bearing will accommodate out of plane motion for the relatively small drifts encountered with the installation. However, bearings can be provided that are capable of up to +/–20° rotation angle.

Schematics of the mounting attachment is provided in Fig 10.7. The mounting pins used to attach the dampers to brackets are often supplied by the damper manufacturer. In most cases, the manufacturer will also provide the brackets used to connect to the structure. The reason for this is that the pins must be fitted very closely to the clevises and spherical bearings, to insure that the connection has no

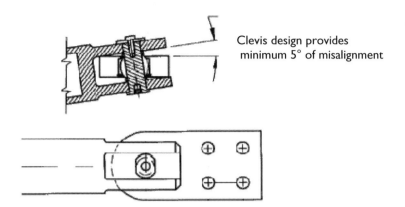

Clevis design provides minimum 5° of misalignment

Figure 10.7 Mounting attachment.

discernible play. The maximum total free play at each bearing is then usually to be 0.01 mm. Note that bolted connection must have tight tolerance holes or a slip critical connection to ensure that the no slippage occurs during continuous dynamic loading.

Note that the engineering analysis should take into account the dynamic stiffness of the dampers and the surrounding elements, especially in applications where the damper forces are high and the deflections are small. This dynamic stiffness accounts for the flexibility of the mechanical elements of the damper, its clevises, the mounting bracket, the mounting pin, and the structure itself. It is important for this to be analysed properly so that the actual deflection and the input velocity are accurately predicted to assure proper motion control of the structure. If the structure or any mechanical element transmitting the damping force is too flexible, this would prevent the damper from stroking properly and therefore absorbing the correct amount of energy.

The resultant dynamic stiffness of the components transmitting the damping force is usually modelled as a simple linear spring in series with the damper.

10.5 MORE DAMPER DESIGN ISSUES

The application of viscous dampers to a footbridge results in additional major design issues, some of which may be unique to the particular structure [3].

One primary issue is to address the fact that the dampers must continuously cycle. It may be understood that that majority of the cycles would take place at low amplitude, but the total number of cycles required by the owner can be based on a 50 year bridge life. For an average frequency of 0.8 Hz, this equates to more than 10^9 cycles of life, far in excess of normal values for any sort of conventional damping device. Ideally, the damper should be maintenance free for the entire life cycle.

Another issue is that the damper must respond to very small deflections as low as 0.025 mm with high resolution. Otherwise the suppression of feedback would not be possible until the footbridge is already well into resonance or under unacceptable motion. Damper frequency response requirements are usually defined as D.C. –20 Hz with a high fidelity output over this entire bandwidth. This issue can be compounded by the fact that due to wind, thermal, and static loadings, total deflections of plus or minus 250 mm or more can be required.

A third issue is that the damper response must have low hysteretic content, to avoid pedestrians sensing the classical "stick-slip" motion of a conventional sliding contact fluid seal, with the resultant perception of instability in the bridge structure. This requirement becomes even more difficult when taken in context with the extremely long cyclic life. This is because conventional hydraulic practice is to use seals with heavy interference for long life under dynamic cycling. These high interferences in turn generate high seal friction, accentuating the "stick-slip" motion.

A fourth issue is that several distinct designs of dampers are often required, each of which require different output forces, deflections, component equations, and envelope dimensions. This is because there might be different modes of vibration to suppress and because of volume, length, or mounting location limitations on the bridge.

A fifth design issue is environmental in nature. The dampers are located outdoors, sometimes over a waterway. The design life is such that all major operating elements of the dampers need to be constructed from inherently corrosion resistant metals that would not degrade over time.

10.6   FRICTIONLESS HERMETIC DAMPER

To address all these additional various design issues, a unique and patented damper can be proposed, previously used exclusively for space based systems. These previous applications have similar requirements for long life and high resolution at low amplitudes, but required relatively low damper forces from small, lightweight design envelopes.

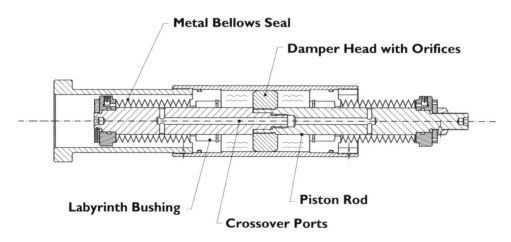

Figure 10.8 Cutaway of frictionless damper.

A cutaway of a typical frictionless hermetic damper is shown in Figure 10.8. The most unique elements of this damper are the frictionless seals made from a welded metal bellows. This type of seal does not slide, but rather flexes without hysteresis as the damper moves. Two metal bellows seals are used to seal fluid in the damper. As the damper moves, the two metal bellows alternately extend and retract, by flexure of the individual bellows segments. Since the seal element elastically flexes rather than slides, seal hysteresis is nearly zero. The volume displaced by the compressing bellows passes through the crossover ports to the extending bellows at the opposite end of the damper. While this is occurring, damping forces are being produced by orifices in the damping head, and the pressures generated are kept isolated from the metal bellows by high restriction hydrodynamic labyrinth bushings. Because hydrodynamic bushings are used, no sliding contact with the piston rod occurs, assuring frictionless performance.

All parts, including the metal bellows, can be designed with low stress levels to provide an endurance life in excess of 2×10^9 cycles. The metal bellows and other moving parts are constructed from stainless steel for corrosion resistance.

10.7 PROJECT EXAMPLES

The Millennium Bridge in London, England was the first application of frictionless hermetic dampers on a footbridge [4]. A total of 37 dampers were constructed, of 7 different types, and are listed in Figure 10.9.

To assure a high resolution output, it is required that all damper attachment clevises be fabricated with fitted spherical bearings and fitted mounting pins, such that zero net end play exists in the attachment brackets.

Testing consisted of three well-regulated crossings on the bridge by a specific crowd at three different walking speeds. A fourth and final crossing was essentially random,

Damper type	Quantity	Description	Use	Stroke (mm)	Length (m)
V1	5	Chevron Damper	Lateral Mode	25	0.7
V2	10	Chevron Damper	Lateral Mode	25	0.7
V3	2	Chevron Damper	Lateral Mode	25	0.7
V4	4	Vertical to Ground Damper	Lateral and Vertical Modes	275	2.3
V5	4	Pier Damper	Lateral and Torsional Modes	60	7.8/3.3
V6	8	Pier Damper	Lateral and Torsional Modes	60	8.2/3.6
V7	4	Pier Damper	Lateral and Torsional Modes	60	8.3/4.6

Figure 10.9 List of dampers for the Millennium Bridge.

Figure 10.10 Pier dampers.

Figure 10.11 Chevron dampers.

with the crowd being told that additional food and refreshments were available at an off-bridge site on a first-come, first-served basis. All of these four final tests proved to be totally anticlimactic – the bridge behaviour being generally described as "rock solid" by the crowd. More importantly to the engineering team, the damped bridge structure performed superbly:

- Peak measured accelerations reduced from 0.25 g undamped to 0.006 g damped.
- Dampers reduced the dynamic response by at least 40 to 1 for all modes.
- No resonance noted of any mode.
- No observable biodynamic feedback occurred.

Damper type	QY	Description	Use	Stroke (mm)	Length (m)
Transverse	I	Transverse damper – BNF side	Lateral Mode	130	1.2
Transverse	I	Transverse damper – Bercy side	Lateral Mode	130	1.2
Longitudinal	I	Longitudinal damper	Vertical Mode	100	0.9

Figure 10.12 List of dampers for the passerelle Simone de Beauvoir.

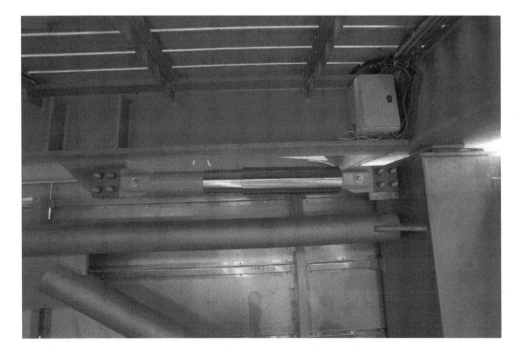

Figure 10.13 Transversal damper, BNF side.

Since then, frictionless dampers have been used for other structures, among others, the Passerelle Simone de Beauvoir connecting the Bercy and Tolbiac sections in Paris.

10.8 CONCLUSION

There are several benefits of using viscous dampers in footbridges. The use of supplemental fluid dampers eliminates the potential for undesirable pedestrian-induced vibrations and biodynamic feedback occurring between the pedestrians and the bridge structure. Viscous dampers allow the use of unique bridge architecture, even when the bridge has modal frequencies which are coincident with normal walking motions of pedestrians. Since the fluid viscous dampers are non tuning-sensitive units, the response of the footbridge can be mitigated in multiple modes.

The application of damping devices to the bridge results in several design issues, some of which can be unique to the studied structure. Therefore fluid dampers are essentially adjusted by the manufacturer to meet specific customer specified parameters.

Frictionless viscous dampers can be proposed for continuous cycling.

REFERENCES

[1] Constantinou M., Symans, M., "Experimental and Analytical Investigation of Seismic Response of Structures with Supplemental Fluid Viscous Dampers", Technical Report NCEER-92-0032, National Center for Earthquake Engineering Research, Buffalo, NY, 1992.
[2] Kasalanati A., Constantinou, M., "Experimental Study of Bridge Elastomeric and Other Isolation and Energy Dissipation Systems with Emphasis on Uplift Prevention and High Velocity Near-Source Seismic Excitation", Technical Report MCEER-99-0004, Multidisciplinary Center for Earthquake Engineering Research, Buffalo, NY., 1999.
[3] Dallard P., et al, "The London Millennium Footbridge", The Structural Engineer, Volume 79, No. 22, pp. 17, 2001.
[4] Taylor D., « Damper Retrofit of the London Millennuim Footbridge, A Case Study in Biodynamic Design », In the Proceedings of the 73rd Shock and Vibration Symposium, San Diego, California, USA, October 26–31, 2003.